Pollution and the use of chemicals in agriculture

Pollution and the use of chemicals in agriculture

Based on the Proceedings of a Symposium held at The Polytechnic of North London, 6-7th April, 1972

Edited by

David E. G. Irvine, BSc, MA, PhD, MIBIOL
Senior Lecturer, Department of Biology and Geology,
The Polytechnic of North London

and

Brian Knights, BSc, MSc, MIBIOL
Lecturer, Department of Life Sciences,
The Polytechnic of Central London

English edition first published in 1974 by
Butterworth & Co (Publishers) Ltd
London: 88 Kingsway, WC2B 6AB

First published in the USA and Canada by
ANN ARBOR SCIENCE Publishers, Inc
P.O. Box 1425
Ann Arbor, Michigan 48106 USA

Library of Congress Catalog Card No 75-92511

ISBN 0 250 40051 0

Typeset by Amos Typesetters
Printed in England by Hazell Watson & Viney Ltd
Aylesbury, Bucks

Preface

Pollution is a highly emotive word at the present time, conjuring up visions of imminent disaster to the environment and the human race, with chemicals synthesised by man himself being the main cause for concern. Such emotional overtones tend to bias any discussion of the subject and the aim of the Symposium with which this book is concerned was to allow, as far as possible, a balanced, rational and up-to-date discussion of one aspect of the chemical pollution problem, that of the use of chemicals in agriculture. These agrochemicals are being used on an increasing scale and concern has been expressed that this will result in the widespread and irrevocable dispersal throughout the biosphere of synthetic, perhaps persistent, accumulative and toxic substances, in the poisoning of living organisms including man and in the disruption of natural ecosystems.

To achieve a balanced discussion, the programme was arranged so that the need for different categories of chemicals by the agricultural industry could be examined and their value in food production assessed. Modes of action of different agrochemicals were considered, which led on to an assessment of their effects on the environment and of how the more dangerous pollution problems could be met by alternative chemicals or practices. The economic and legislative aspects, the controls and safeguards exercised by government and industry were also considered. Contributors were chosen to represent farmers, the agrochemical industry, government organisations and educational institutions. Thus, it was hoped, the views of both the economically orientated agriculturalist and the ethically and aesthetically orientated environmentalist could be given equal airing, and rational and balanced conclusions reached.

Whether the Symposium achieved these aims can only be judged in hindsight when this volume is read in five or ten years time. It could at least be said that the two possible extremes were avoided; there were no harangues by opponents of our synthetic and technologically-based society seeking a return to an idyllic but mythical Golden Age of 'natural' farming methods, nor were there whitewashing exercises glossing over real difficulties or malpractices.

Contributors brought out the fundamental fact that pollution and the cost of combating it are to be balanced against the need to provide more and better food for the world's expanding population.

One conclusion agreed upon in discussions was that there was a lack of communication between the agrochemical producers and users and the environmentalists, both professional and amateur. The mass media were thought to be particularly to blame for this because of their tendency to give publicity to one-sided and gloomy prophecies. While there is no easy answer to this communication problem, the Editors hope that the publication of this volume will help; at best, we think it gives a balanced view of the problem, at worst it puts forward views and facts generally given least publicity and will, we hope, enable more balanced discussions to be held.

D. E. G. IRVINE
B. KNIGHTS

Acknowledgements

The Editors would like to record their gratitude to Mr. D. Etherington, Head of the Department of Biology and Geology, for his help and encouragement in the arranging of the Symposium. Thanks are also due to his staff, in particular Dr. K. Parry, and that of other Departments in The Polytechnic of North London for administrative and technical assistance, especially to the staff of the Electronics and Telecommunications Department for assistance with audio-visual aids.

It only remains to thank individual contributors and those attending the Symposium for making it so stimulating and thought-provoking and Butterworths for their interest in making these contributions available to a wider audience.

Contents

One

The need for chemical control of pests and the use of fertilisers

H. C. Mason
Technical Director, National Farmers' Union

SUMMARY

Food is essential to life, and the historical development of farming has led from subsistence systems to the commercial systems that are a necessary prerequisite for industrialisation. One of the biggest problems in the world today is the population explosion, aided as it has been by advances in medical knowledge. Equal effort must be put into ensuring that enough food is available to feed this increasing population and this can really only be achieved by increasing productivity. There is little potentially good farmland left in the world for reclamation and even existing good farmland is being lost to increasing urban, industrial, amenity and conservation demands. Maximum crop growth can only be obtained when nutrients are added or returned to the soil. The advantages of chemical fertiliser usage over other methods such as recycling of manure and sewage sludges are discussed, as are the means of increasing productivity by chemical control of animal pests, fungi and weeds. The present inadequacies of both biological control methods and breeding of resistant crop varieties are discussed. It is concluded that agrochemicals, responsibly used as they are, are an essential adjunct to the modern farming methods needed to produce enough food of good quality to feed the world's population.

INTRODUCTION

It is worth remembering at the outset a fundamental fact that food, the availability of which is taken for granted in this country, is essential to life and that therefore farming, the business of producing food, is also essential. For millions of people throughout the world who have either an inadequate supply of food or have deficiencies in their diet, starvation and malnutrition are very real problems.

In ancient times man was originally a hunter, depending on the availability of wild animals and his ability to catch them while supplementing his diet with wild fruits and other naturally occurring foodstuffs. Later he turned to the more reliable methods of domesticating animals and growing his own crops. Some of the earliest agriculture consisted of shifting systems of farming where an area of forest or scrubland was burnt and the fertility resulting from the ashes used to grow crops for a year or two. When the fertility was exhausted, the farming community moved on to a new area—a system rather devastating to the natural environment. Later, more settled systems of farming were developed but these were still essentially subsistence systems, and, indeed, it is worth noting that farming is today still the occupation of the majority of people in the world as a whole. It is only as farming becomes more productive and produces a surplus that can be sold off the farm on a commercial basis, that people can be released from subsistence farming to follow other occupations, thus increasing the scope for technological and industrial development. The greater the efficiency of agriculture in any country, therefore, the greater the scope for industrialisation. It is interesting in this respect to examine the proportion of the population engaged in farming in different countries (see *Table 1.1*). From the figures given, it is obvious that

Table 1.1 THE PROPORTION OF THE POPULATION IN DIFFERENT COUNTRIES ENGAGED IN FARMING (OECD, 1971)

Country	*Percentage of the population engaged in farming*
Turkey	70
Italy	21
France	15
Netherlands	7.5
USA	4.5
UK	3

there is, on the whole, a fairly straightforward relationship between the degree of industrial development and the efficiency of farming and the percentage of the population engaged in farming. Every country does not, of course, produce all the food that it consumes—the UK only produces half the food it needs, for example. Even with double the present farming population, however, it would still only be 6 per cent of the total population, a very low figure compared with most other countries.

One of the biggest problems facing the world today is the increase in population. It has been estimated that about 2000 years ago there were no more than 250 million people in the world, but that this figure had doubled by the seventeenth century, doubled again by the mid-eighteenth century and doubled again by 1930 so that by then the estimated world population was about 2000 million. One estimate for the present population is 3500 million and, according to some calculations, this may be doubled again by the end of this century, every increase meaning more mouths to feed. Malthus predicted in the nineteenth century that population increases would bring disaster because of limited food supplies. His predictions have been largely discredited because he failed to foresee the productivity of modern farming and the scope for extending the area of farmed land—but the principles on which he based his thesis are still quite sound. Clearly populations cannot go on increasing indefinitely unless food supplies can be equally well increased. Medicine and hygiene have done a great deal to promote the increase in population, reducing infant mortality, controlling debilitating diseases such as malaria and bilharzia and enabling people to live longer; it must be ensured that science contributes equally with aid to food producers in order that enough food can be provided for these people.

Increased food production can be achieved in two different ways; by increasing the area of land devoted to farming or by increasing the production on the land already farmed. The former is not easy; considering the total surface of the world, the greater part is water and much of the land area is climatically unsuitable, for example in polar regions and in deserts and mountains. The most suitable land has already been used for farming and even the area of this land is being reduced by urban demands for more houses and roads, more industrial development and more amenity areas, while more demands are now being made for conservation areas such as nature reserves. A study of the position in this country reveals that something in the order of 50 000 acres of farmland are being lost every year to the sort of demands listed above and reveals that something like one and a half million acres have been lost since the last war, usually best lowland farming land. Although farming has been reclaiming marginal land, this is of less productive value. Farmers are trying to produce more and more food from fewer and fewer acres with less and less manpower and this can only be done with the help of fertilisers, pesticides and mechanisation.

THE NEED FOR THE USE OF FERTILISERS

In relation to the growing of plants, one of the main requirements is carbon dioxide in the air. Carbon dioxide is not normally looked on as a fertiliser since this factor cannot be controlled for field crops, although of course in enclosed glasshouses the concentration of carbon dioxide is often artificially raised in order to increase plant growth. Other nutrients required for plant growth are nitrogen, potassium, calcium, magnesium and sulphur. These are the major elements taken up from the soil, together with the trace elements, manganese, iron, boron, copper and molybdenum. Shortage of any one of these will prevent maximum growth from being obtained. They are usually present in the soil but in varying degrees of solubility and availability. Weathering of mineral particles gradually releases more nutrients in the soil, nitrifying bacteria help by adding nitrogen to the soil and some nutrients are returned in decaying vegetation. There is also a constant loss, however, in leaching from the soil through drainage and also in removal through cropping, whether this be in removal of cultivated crops such as, for example, harvested grain, or in the cropping of grassland by grazing animals. Removal of nutrients from the soil without replenishment means that the soil will be impoverished which in turn will lead to low yields.

There is nothing new in returning nutrients to the soil. Sanskrit writings of 3000 years ago refer to the value of dung and the Romans recorded the use of lime and also the value of legumes in adding to soil fertility. In the twelfth century there were references to the growth-promoting effects of ash, and bones were certainly used as a fertiliser by 1650. One of the most interesting historical records is that of one Francis Home, in 1757, who proved that nitrates, ammonium salts and also potash salts were beneficial in furthering the growth of plants. It was assumed, however, that even if they were beneficial, they would be far too expensive to be used as fertilisers (OEEC, 1957). Guano was first shipped to Europe in 1810 and during the next 50 years Chilean nitrates were introduced and superphosphate and potash fertilisers were produced. Therefore all three major elements—nitrogen, phosphorus and potassium—were available as fertilisers 100 years ago. It was not, however, until the synthetic nitrogen industry developed at the beginning of this century, and expanded so much for fertiliser production after World War I, that modern large scale fertiliser usage became possible.

Before looking at fertilisers in more detail, it is worth considering alternative means of adding nutrients to the soil. Dung or farmyard manure returns some of the elements removed in cropping but recycles only a proportion of the nutrients since only some crops are used for livestock feed while there is also an inevitable loss in drainage. It is true that the recycle value can be enhanced by any bought-in feedstuffs which leave their residues in the manure. Indeed, at one time it was a popular practice to feed excessive quantities of linseed cake to fattening cattle, not only for the bloom that it put on the cattle, but also because of the high nitrogen levels it left in the manure—a quite uneconomical process, needless to say. Buying extra manure can help, but at best only results in transferring fertility from one farm to another, and, in the case of bought-in feedstuffs, perhaps from one country to another. Farmyard manure is very bulky and difficult to handle and it is worth noting that some 10 tons (10.2 tonnes) of manure are equivalent in fertilising value to only about 3 cwt (152.4 kg) of modern fertiliser. Thus, although very useful, it can be seen that it has its limitations for the purposes of increasing field fertility. Another possibility for consideration is the use of sewage sludge because, after all, a great deal of the food produced is eaten by human beings and use of sewage sludges can therefore be regarded as an extension of the use of the system of recycling nutrients on the farm through the use of farmyard manure. In practice, however, there are many problems involved, because sewage sludges are not produced in bulk in farming areas and transport costs are very high, making the whole business uneconomical. Also, there is a very real and serious danger of pollution of the land because, surprisingly, sewage sludges contain residues of quite a number of heavy metals—zinc, copper, chromium, nickel and lead—which can build up to toxic levels in the soil (ADAS, 1971). There are several instances where land has already been made sterile as a result of such sewage pollution. It has been maintained that such elements are only found in sludges from industrial areas where they might originate from metallurgy or other industrial processing plant. It has been found, however, that zinc, which is of particular concern, is present in almost all sewage sludges although its origin does not seem quite decided. It has been suggested that it comes from galvanised guttering and water tanks, from zinc which is used to put a shine on toilet paper and also a great deal of zinc is used in cosmetics.

Nitrogen fixation by bacteria in the soil is another useful way of adding to soil fertility. These bacteria are encouraged by good arable cultivation and there are also bacteria which work in symbiosis with leguminous plants, and these can contribute significant amounts of nitrogen for plant growth but, in temperate climates, cannot supply all that is needed. Some countries are more fortunate in having longer and warmer summers and the bacteria can continue nitrogen fixation for longer periods. The most that farmers in the UK are likely to benefit from the activity of symbiotic bacteria is in the order of 80 units of nitrogen per acre (1 unit = 1.12 lb (0.51 kg)) when growing a leguminous crop. Unfortunately, the majority of crops that need to be grown in this country are not leguminous. This leads back, therefore, to the conclusion that the only way in which the fertility of soils can be appreciably added to is through the use of so-called 'chemical' fertilisers.

Long-term fertiliser trials were started at Rothamsted Experimental Station in the 1840s and from 1852, in the famous Broadbalk Field, wheat has been grown continuously with the same manuring on the same plot each year, until in 1935 a fallow cycle was introduced for the purposes of weed control. On a plot receiving no manure at all, the yield fell off very quickly in the first few years and then continued to decline at the rate of about 0.5 cwt (25.4 kg) every 10 years as the soil gradually became more impoverished of nutrients (Rothamsted Experimental Station, 1969). The yield is now at what is considered in this country a hopelessly uneconomic level but is nevertheless about the world average for wheat. On the other hand, the manured plot receiving 14 tons of farmyard manure an acre (5.7 tonnes/ha) each year continued to give yields near the maximum, being slightly outyielded by a plot receiving a complete chemical fertiliser. This occurred even though the fertiliser plot received less nitrogen and potash in total. Over the period 1852–1967, the fertilised plots gave an average increased yield of 175 per cent over the unmanured.

There are plenty of other field trials that have been carried out in the UK to show that there is an almost linear gain in yields in response to increasing rates of nitrogen application, provided the other essential elements are present, until the law of diminishing returns begins to apply once optimum levels have been reached. It is not possible, however, to do a simple sum to calculate increases in yields from adding different amounts of fertiliser; soil types in this country are so variable, there are so many levels of fertility and there are other confusing factors such as pests, variable weather conditions and different cultivation methods. It is also not possible to say exactly to what extent the enormous post-war increases in yields are due solely to the use of fertilisers. Various experiments could be quoted to show dramatic increases but these involve the choice of a field with low initial fertility to which the exact nutrient requirements are subsequently applied—this does not give a really true or typical picture. Dr. George Cooke, however, has made a study (Cooke, 1960) of field experiments throughout the country and has calculated the average optimum dressings for food plants over the country as a whole for the major crops. This gives some interesting estimations of averages; for example, he calculated that extra yields to be expected from optimum fertiliser dressing would be 5 tons per acre (2.05 tonnes/ha) for potatoes and 7.5 cwt per acre (153.9 kg/ha) for cereals. The former represented something in the order of a 40–50 per cent increase in yield, the latter about a 25 per cent increase. It is obvious from these figures why farmers continue to use fertilisers—they would not do so if they did not see that they were getting increased yields and economic returns. Fertiliser usage has been increasing steadily in this country, the use of nitrogen (expressed as tons of elemental nitrogen) showing a particularly large increase since 1939 (see *Table 1.2*).

Table 1.2 THE AMOUNTS OF DIFFERENT FERTILISERS APPLIED ON FIELDS IN THE UK IN 1939, 1949 AND 1969 (OEEC, 1957; FERTILISER MANUFACTURERS ASSOCIATION, 1970)

Year	*Tons of fertiliser applied per year*		
	Elemental nitrogen	*Phosphate* (P_2O_5)	*Potassium* (K_2O)
1939	60 000	170 000	75 000
1949	185 000	419 000	196 000
1969	782 000	460 000	430 000

In relation to phosphates, a point is being reached in the UK where many crops are receiving about the optimum level of application. The greater use of fertilisers, together with the use of pest control products and new crop varieties, has been largely responsible for the striking increases in yields achieved in British farming. Usage at comparable levels is confined mostly to Western Europe and North America, but in the world as a whole farming is less intensive and less productive with yields nothing like as high.

THE NEED FOR THE USE OF PESTICIDES

The supply of nutrients to plants is only part of the story, however, since specially bred varieties of plants selected for their high quality and high yield characteristics are now being grown. It is also essential that the plants are protected from competition with other plant species and are protected from attack by parasitic fungi and by animal pests such as, for example, rodents, birds, insects, eelworms and slugs. It is only if these pests can be eradicated and all the nutrients required can be supplied that maximum yields can be obtained. The fact that yields have been increasing is an indication that we are gaining on some of these rivals. History is in fact studded with reports of famines and devastations caused by pests such as locusts, and an example in the British Isles was the famine towards the middle of the last century when the potato crop failed because of attacks of potato blight fungus over a period of several years. It was the cause of over a million deaths in Ireland and without crop protection products, this country could be faced with similar devastation in an unsuitable season. It is unlikely, of course, that people would face starvation, because food could be imported from other countries, but nevertheless much hardship would result, so there is a real need to protect against such pests. The potato blight fungus grows on the foliage in warm, damp conditions and kills it off, inhibiting further growth, but if unchecked also spreads down to the tubers. This causes them to rot either before they are lifted or when in store and, of course, rots the seed potatoes for the succeeding crop. Nowadays protective fungicides are in regular use which, when sprayed on the leaves, prevent the establishment of the fungus, and it is estimated that 70 per cent of our national crop is protected in this way.

There are so many pests and diseases that can attack crop plants that it would not be worth while in the space available to list them all or describe the symptoms and trouble that they cause. There are the mildews and rusts and other leaf diseases of cereals which are causing increasing problems, while others such as smuts on cereals are fairly well controlled by the use of fungicidal seed dressings, a practice which started in the 1930s. Nowadays, protective seed dressings can also include insecticides which will give protection against the wireworm and wheat-bulb fly. Wireworms have caused many complete failures of cereal crops, particularly when the ploughing-up of grass campaign started in the 1940s, but when sampling techniques were evolved it was possible to judge which fields had too many wireworms to make it safe to grow cereal crops. At that time suitable chemicals were not available to give the requisite protection. Some people say that such things can be left to nature and that birds ought to be able to deal with wireworms, but it is known from personal experience that a field of barley was completely destroyed by wireworms; while, at the same time, the birds were busy eating a crop of maize that had been planted on another field on the farm. The end result was two completely bare fields!

It is not, however, just a question of protecting a crop against complete destruction, the quality of the crop has also to be considered. Potatoes for instance can also be attacked by wireworms which just burrow their way through the potatoes, leaving unsightly marks. No one these days wants potatoes like this, anymore than wanting scabby apples or apples with grubs in them. Quality of produce is also very much a matter of protecting against pests.

Some insect pests, such as aphids, not only cause direct damage to crops but indirectly cause a great deal more trouble by spreading virus diseases which very considerably reduce yields of crops such as potatoes and. sugar beet. In the absence of protective spraying, virus yellow can be a serious problem in sugar beet. If spraying is carried out when there are warnings that aphids are building up to serious proportions, it is calculated that an increased yield is obtained of on average an extra ton per acre of sugar beet. It is impossible in a limited space, as pointed out above, to list all the pest problems that farmers experience, and if one were to look at the full list it is perhaps amazing that farmers ever manage to grow crops at all. Equally there are external parasites of livestock, such as lice and blowflies, and internal parasites such as worms, all of which have to be controlled.

On examining the question of the fight against pests, it is found to be as ancient as farming itself. The earliest record of the use of fungicides is by Homer some 3000 years ago who referred to the pest-averting properties of sulphur, which is still used as the basis of some modern fungicides. There are various other methods of dealing with pests; for example, infected plants can sometimes be removed or insect pests picked off by hand or trapped. The flea-beetle used to be a considerable problem on Brassica crops and little could be done about it. One method of control sometimes recommended was to cross the fields with harrows or drills from which dangled tarred sacks so that as the flea-beetles jumped up, they stuck to the sacks. Fly papers may still be a very good means of keeping domestic fly populations down, but on a field scale, particularly under dusty conditions, it is not a very practical method. Nowadays such problems can be avoided by using dressings on Brassica seed which ensure that the numbers of flea-beetles do not build up.

Cultural controls can also be sometimes used against pests. Long rotations of crops can reduce the opportunities for pests to build up because there is then no food for them in intervening years between the particular crops on which they feed. The potato-root eelworm is an example of this; in a long rotation, many of the eelworm cysts remaining in the soil die off between the growing of one potato crop and the next. Even after nine or ten years, however, some cysts still survive, ready to cause another infestation. The wheat-bulb fly also deserves a brief mention, since land left bare in August is an invitation to the flies to lay eggs there. Often in farming practice wheat is sown after a fallow, and so grubs hatch out from eggs laid on the bare ground, attack the subsequently sown wheat and cause devastation. By ensuring that ground is not left bare in August, a lot of the trouble from bulb fly can be avoided. Sugar beet can be planted, for instance, but in a dry season it may not cover the ground sufficiently by August and bulb fly damage could then still occur.

Pest problems can sometimes be avoided by legal measures. The Colorado beetle, for instance, is not found regularly in this country but is a problem in many others. Controls over the imports of plants (plant health regulations) prevent the beetle from being introduced into this country and so regular protective spraying programmes are not needed.

The idea of breeding crop varieties resistant to pests such as fungi and insects is very attractive. There are not very many varieties available that are resistant to insects, however, but if varieties resistant to all pests could be found, it would be a very much easier world for the farmer. Wart disease of potatoes is less of a worry these days because there are good resistant varieties, but plant breeders have been working for years to produce potatoes resistant to blight. They have found some varieties which show reduced susceptibility but as fast as they make progress in breeding new varieties, the blight fungus seems to be making equally fast progress in breeding new strains which can attack the new varieties. Similar problems have been found with cereal rusts and there are many other examples where breeding of effectively resistant plants is similarly limited.

Farmers would like to see very many more cases where biological control can be practised, but although success stories are heard from overseas, few such methods have been found that can effectively be used in this country. With very many of the pests now controlled biologically, the problem arose originally when, for one reason or another, the pest was introduced into a country where its natural predators were lacking. When predators have been introduced later, this has produced startling successes in reducing pest numbers. Some biological control is now available for use in glasshouses, where of course the environment is fairly well controlled. There is a great need for further research on biological control measures and this is something all those concerned with agriculture would very much welcome. But one has to remember that it is usually the pest that is controlling the predator in effect, since if the predator does its work too well it will run short of food and will then die off.

It is hoped that more subtle methods of pest control can be introduced, such as the technique of releasing sterilised male insects which mate ineffectively, thus reducing the birth rate of the population. This has exciting possibilities but is unfortunately of little use with promiscuous species or where distribution is too widespread and not enough males can be sterilised, released and mated with females to affect population numbers significantly. The use of sex attractants and other lures has possibilities, but at present these are very much in the theoretical stage and have not been developed for practical use in this country. Again, at the present time, such methods do not offer an efficient alternative to control of pests by chemical means.

THE NEED FOR THE USE OF HERBICIDES

Weeds are another serious problem, since they compete with crop plants for daylight, nutrients and moisture. There are always far more weed seeds in the soil than seeds of the crop being grown. At the Weed Research Organisation some sampling and calculations showed that for one weed species alone, there were more than 100 million seeds per acre (ARC, 1965), and of course weeds multiply at a tremendous rate. It has been calculated, for example, that since chickweed has several generations in a year and each plant produces 2000 seeds or more, one seed could multiply to as many as 2000 million by the end of a season. Farming has always been a battle against weeds. In the olden days this was tackled mainly by fallowing, a system where no crops were grown for a year, with frequent cultivation to kill all weeds as they grew. In cropping rotations practised in this country until the nineteenth century, one third of all the arable land was fallow at any time. Even with the introduction of root crops grown in rows and regularly hoed, fallows persisted until the time of the agricultural revolution after the last war, but nowadays it is a very rare practice. Hoeing is not always effective under wet conditions, and during dry weather it deprives the soil of valuable moisture. Also, there is always a certain amount of damage to the roots of crops. Today, workers are just not available for such tasks in this country; the work force in the industry has been halved and weed control rotations require too rigid a system of farming to be economic or to produce the crops required.

In the 1930s, control of weeds such as charlock was attempted using scorching agents like kainite or cyanamide or other products such as sulphuric acid and later DNOC (dinitro-orthocresol). The first modern chemical herbicides used were phenoxy-acetic acids which have been available since 1945, providing an almost magical degree of selective weed control in cereal crops, and indeed opening the way to continuous cereal growing. Since then over a hundred different chemicals with herbicidal properties have been described and an extensive range of products is now available to control different weed species in a wide variety of crops (British Crop Protection Council, 1968). These chemicals may be selective or non-selective, may be used for pre-sowing, pre- or post-emergence and may kill by contact, translocation or residual action. Products are also available that will kill foliage so that no ploughing or cultivation is needed before another crop is sown, saving labour, time and soil moisture. The use of herbicides is a complicated business, based on a whole new science of weed control which has evolved within the past 25 years.

CONCLUSIONS

Some general examples of the need for the use of fertilisers and crop protection products have now been outlined and some of the drawbacks associated with methods not involving the use of chemicals have been discussed. Expenditure on agrochemicals is quite high; something in the order of £184 million was spent on fertilisers and nearly £20 million on crop protection problems in this country in 1971 (Annual Review and Determination of Guarantees, 1972; ADAS, 1971). (The greater part of the latter, was spent on herbicides and it should be noted that these have on the whole a very low mammalian toxicity.) It is now perhaps appropriate to summarise the advantages of the use of agrochemicals in terms of the overall effects on agricultural production, and wheat can be taken as a good example. New varieties of wheat, as with other crops, have helped to increase yields, but it has to be remembered that these new varieties are ones which require conditions of high fertility to produce the increased yields and therefore their introduction has also encouraged and made possible the use of more fertiliser. If the older varieties were given too much nitrogen, laid crops resulted which could not be harvested; today's varieties are able to utilise more nitrogen and give higher yields of grain. The present world average yield of wheat is of the order of 12 cwt per acre (236.2 kg/ha). The average yield in this country in 1946–47 was just under 1 ton per acre (0.41 tonnes/ha); today it has risen to 1.6 tons per acre (0.66 tonnes/ha). Not so long ago, a crop of wheat which it was estimated would yield 2 tons per acre was thought to be something the like of which would never be seen again. Today crops of 2 tons per acre are far from remarkable and indeed it seems likely that before very long the national average yield of wheat will be of this order. Already a yield of 3.5 tons per acre has been produced from one field, and other crops could be quoted to show similar achievements. Over the past 25 years the output from the agricultural industry as a whole has more than doubled, so that today two-thirds of this country's temperate grown food is home grown. If the clock were turned back with regard to production methods, the UK would be faced with importing half the present output from the agricultural industry, i.e. half of an annual output worth £2800 million at present. Assuming the food was available to be imported, and assuming also that world prices would not be affected by this increased demand, one can imagine what effect this extra expenditure would have on this country's delicately poised balance of payments situation. Responsibly used, as indeed they are in Britain, fertilisers and crop protection chemicals are giving immense benefits to all, and are essential requisites to modern farming.

REFERENCES

Agricultural Development and Advisory Service (1971). *Advisory Paper No.* 10. *Permissible levels of toxic metals in sewage used on agricultural land.* London; Ministry of Agriculture, Fisheries and Food

Agricultural Research Council, Weed Research Organisation (1965). *First Report 1960–1964*, Oxford; Weed Research Organisation

Annual Review and Determination of Guarantees (1972). (Cmnd 4928), London; HMSO

British Crop Protection Council (1968). *Weed control handbook*, Ed. J. D. Fryer and S. A. Evans. 5th edn. Vols. 1 and 2, Blackwell Scientific Publications, Oxford

Cooke, G. W. (1960). *Fertilisers and profitable farming*, Crosby Lockwood

Fertiliser Manufacturers' Association (1970). *Fertiliser Statistics 1969*, Fertiliser Manuf. Assoc.

Rothamsted Experimental Station (1969). *Report for 1968, Part 2: the Broadbalk Wheat Experiment*, Rothamsted Exp. Stn.

OECD (1971). *OECD Observer, February 1971.* OECD

Organization for European Economic Co-operation. European Productivity Agency (1957). *The efficient use of fertilizers including lime: Report on a Training Course held in Denmark 31 May to 24 June 1954*, OEEC

DISCUSSION

Sheen: *Ministry of Agriculture, Fisheries and Food.* Do you attach any significance to the theory held by some people that continual heavy dressings of artificial fertilisers may be detrimental to soil structure?

Mason: There has recently been a very detailed study of soil structure and fertility, the results of which indicate that farmers have not always been as considerate to their soils as they might have been, very often through using arable cultivation on soils not naturally suited to that purpose. There has, therefore, been more fertiliser usage on these soils but there is no evidence that this is in itself harmful to soil structure.

Clampitt: *Cooper Technical Bureau.* Surely there are some problems associated with the use of artificial fertilisers, since phosphates are believed to be concerned with eutrophication and nitrates are coming under a certain amount of suspicion because of the formation of nitrosamines which are carcinogenic.

Mason: It is not part of my brief to deal with pollution but I am not going to dodge this question! Phosphates become fixed in the soil and I know of no evidence for any significant phosphate run-off from soil. There are certainly phosphates to be found in many river waters but this is largely due to phosphates used in detergents which have found their way back into the river systems. As for nitrates, there may be instances where these find their way into our rivers but these may equally well come from soils which are not getting any fertiliser at all, nitrates here being produced by bacteria. This goes on in small proportions all the time. Again, the nitrate problem is not particularly one arising from the use of fertilisers; in nearly every case where high nitrogen levels are found in rivers, this is due to sewage outfall. As far as nitrates and nitrites in plants are concerned, it is true that some plants, for example spinach and mangolds, do contain these. This is a natural occurrence. It is a question therefore of choosing foodstuffs to suit your particular animals, a fact appreciated by farmers.

Knights: How well educated are farmers themselves in the complexities of the overall agricultural and agrochemical situation you have discussed?

Mason: It is impossible for everyone to be fully conversant with the problems. The main principles of NPK fertilisers, herbicide usage, and so on are well understood but it is necessary that a full team of experts be available to give detailed technical advice where appropriate.

Griffin: *University of Maine.* Would it be possible to hazard an estimate of the approximate amounts of animal manures recycled back onto the land as fertiliser as opposed to any other means of disposal in the UK.

Mason: The vast majority of animal manure is recycled but in the case of certain intensive systems, for example pigs and poultry, there may not be sufficient land available on the farm for recycling. In many such cases, arrangements are made with neighbouring farmers to take the benefit but a small proportion of animal manure is wasted because of disposal problems. Personally, I would guess that about 90 per cent of manure is recycled.

Warrick: *Greater London Council.* Do you think farmers fully appreciate the harmful effects that hydrocarbon insecticides such as DDT can have on insect-eating bird populations?

Mason: In relation to the direct effects of such insecticides on birds, there have been scares but the problem has probably been over-exaggerated. There is circumstantial evidence that birds of prey have been affected by insecticides but it has not been shown conclusively that they have been killed off in any large quantities this way, certainly nothing in proportion to those killed, for example, on roads or around lighthouses. Nobody is going to condone the fact that insecticides may sometimes cause losses of certain birds but I believe that the majority of losses have been due to loss of suitable environment. Agriculture and the use of agrochemicals are bound to upset the balance of nature, that is what farming is all about. The answer perhaps is to intensify our farming and limit the acreage under cultivation so that we can have more conservation areas completely undisturbed. This is probably a good reason for using more agrochemicals.

Puri: *Liverpool Polytechnic.* Nearly 80–85 per cent of the world's fertilisers are used by the developed countries, very little being available to the developing countries because of cost. Will they have to depend on other means of increasing productivity—new varieties, biological control, and so on—or would it be possible for the developed countries to release some fertilisers to those countries where conditions are better for producing larger yields? Tropical forests could be particularly important areas where application of small amounts of fertiliser would give much larger yields than in some temperate areas.

Also what do you think will be the world's total fertiliser needs if populations go on expanding at the same rate and will we be able to meet them by the year A.D. 2000?

Mason: I do not think there should be any monopoly of fertiliser use in developed countries. I believe part of the development process of other countries must be to build their own nitrogen-fixing plant to produce their own nitrogenous fertilisers. Fertilisers are bulky materials to transport around the world so it is up to every country to produce as much as it can itself. In relation to naturally occurring products such as phosphates and potash, these may have to be transported from one country to another. There are immense reserves of these materials and I think there is no more fear that we shall run out of these in the immediately foreseeable future than that we are likely to run out of our supplies of, say for example, energy in fuel oil.

Two

Some mechanisms of chemical control of insect pests

R. C. Reay
Department of Biological Sciences, Portsmouth Polytechnic

SUMMARY

A number of insecticide groups are discussed with reference to the nature and site of their biochemical lesion in target organisms, especially where this is found in the nervous system. Rotenone and the methylenedioxyphenyl synergists are quoted as examples of biologically active compounds which affect other systems. The modes of action of insect moulting and juvenile hormones and their mimics are then discussed, these being considered as typical examples of areas of recent advances in chemical pest control techniques. Possible relationships between activity against target organisms and effects on non-target species are briefly considered.

INTRODUCTION

The subject matter of this paper is presented under two broad headings which reflect different stages in the evolution of man's efforts to control insect pests by chemical means. Modern pesticide technology dates from the late 1930s and since that time there has appeared a vast range of compounds (insecticides) which kill pests (target organisms) by poisoning them. These 'second generation' pesticides (Williams, 1967) are synthetic organic compounds such as DDT and have largely replaced an earlier generation of materials based on inorganic heavy metal derivatives. The latter originally fell out of favour because of their lack of selectivity and now some of the second generation pesticides are known to present problems both by being environmental contaminants and by the fact that they select strains of resistant pests. These limitations have stimulated research into other chemical control methods and during the past decade a third generation of materials has emerged which together represent a more varied approach to the pest control problem. Many are not at all toxic in the accepted sense, and some actually require target insects to remain alive for some time after treatment in order to work effectively. They are relevant here because they could represent the next generation of problems which face those concerned with environmental pollution.

For the present purpose, a pesticide of any sort becomes a pollutant when it adversely affects any organism (including man) other than the target organism. This is undoubtedly a broad definition, but it should be emphasised that the apparent absence of side effects is no reason for automatically assuming that the compound involved is safe. The fact that damage to the ecosystem may be long term and unpredictable should counsel caution.

MODES OF ACTION OF CONVENTIONAL INSECTICIDES

Although there is often a complex set of symptoms associated with insecticide poisoning, it is usually possible to trace them back to one important reaction within the organism. This takes place at what is known as the site of the biochemical lesion and it is what happens here, rather than general symptoms, with which this chapter is concerned. Because a number of insecticides which have been implicated in environmental pollution are thought to produce lesions in the nervous system of both target and non-target organisms, the functioning of this will be briefly examined.

Transmission of a nerve impulse *along* a nerve axon is electrochemical, and is associated with precise and ordered changes in the concentration of K^+ and Na^+ within and without the nerve cell membrane as the relative permeability of the membrane to those ions changes in a controlled fashion. Transmission *across* the junction (synapse) between adjacent nerve cells or between a nerve cell and a muscle is mainly 'chemical'. When an impulse reaches a synapse it triggers the release of a chemical which diffuses across the intervening space and induces an appropriate response in the next cell, again by altering membrane permeability to sodium and potassium ions. The main interest here is in the system where acetylcholine (ACh) is the neurotransmitter, because this has been the one most frequently investigated. But it should be realised that the nerve–muscle synapses of insects are not of this type: various other chemicals have been implicated in this group of organisms. In order to restore the sensitivity of the synapse, ACh must be destroyed and this is brought about by the enzyme acetylcholinesterase

(AChE) which splits it into acetic acid and choline. Albert (1968) quotes data which show that in some vertebrates 5×10^6 molecules of ACh are released by a nerve terminal in one impulse: however, enough AChE is available to split two hundred times this amount every millisecond.

With certain qualifications, many of the symptoms of organochlorine poisoning are in accordance with what one would expect from a disruption of axonic transmission. Thus DDT (dichlorodiphenyltrichlorethane) produces hyperactivity and then fatal paralysis in treated insects. As long ago as 1955, Mullins (1955) suggested that variations in the toxicity of the isomers of BHC (benzene hexachloride: hexachlorocyclohexane) might be related to the actual *shape* of the insecticide molecule. His theory required the presence of pores of a particular shape and size in the nerve membrane (i.e. that it comprised a lattice structure) into which active isomers would fit more tightly than inactive isomers. More recently, Holan (1969) has postulated a similar situation for DDT. Because DDT is lipid soluble it is suggested that the molecules become distributed at the protein–lipid interface of the nerve membrane, with the phenyl ring part of the molecule locked on to the overlying protein. At the same time the trichlorethane moiety acts as a wedge to keep open some site, such as a pore channel, in the lattice which allows leakage of Na^+, so disrupting the normal Na^+/K^+ ratio and resulting in uncontrolled nerve activity.

Less is known about the mode of action of the cyclodiene type of organochlorine insecticides such as dieldrin, despite the fact that they are widely suspected of being environmental contaminants. There is some evidence that they disrupt axonic transmission, but whether the site of the biochemical lesion is in the nerve membrane or not is open to speculation.

Although they bear no chemical relationship to the organochlorines, the pyrethroid insecticides also appear to act as nerve poisons, possibly by changing nerve membrane permeabilities. They are mentioned here because natural pyrethrins have been widely used for many years as household insecticides while, more recently, Elliott (1971) has described a new range of synthetic materials some of which are now being produced commercially and will eventually find their way into the ecosystem.

Organophosphorus insecticides (OPs) and carbamates in contrast to the above groups affect synaptic transmission. In general such compounds do not persist in the environment as long as the organochlorines but they have been used on a vast scale and must be taken seriously as potential pollutants. It is thought that when an OP reaches its site of action it initially *combines* with AChE to form a complex and then reacts *chemically* to form a stable phosphorylated enzyme. This reaction probably takes place between the phosphorus atom of the toxicant and the — OH group of the amino acid serine which comprises a vital part of the enzyme protein. The phosphorylated enzyme so produced is unable to catalyse the hydrolysis of ACh, which accumulates at the synapse producing excessive and uncoordinated activity and eventually death. There is also evidence that the spatial arrangement of some OPs may bear some resemblance to that of the ACh molecule and that this, as well as reactivity, is responsible for inhibition. One further point can be made about the mode of action of OPs; a number are converted *within* organisms by their own metabolic processes to more potent AChE inhibitors. This is known as activation and the net result is that such compounds arrive at the site of the biochemical lesion in a more potent form than that in which they were applied. This is often due to the conversion of the P=S group to a P=O, one well-known example being the conversion of parathion-methyl to paraoxon as shown in *Figure 2.1.* Such 'oxidation' products are

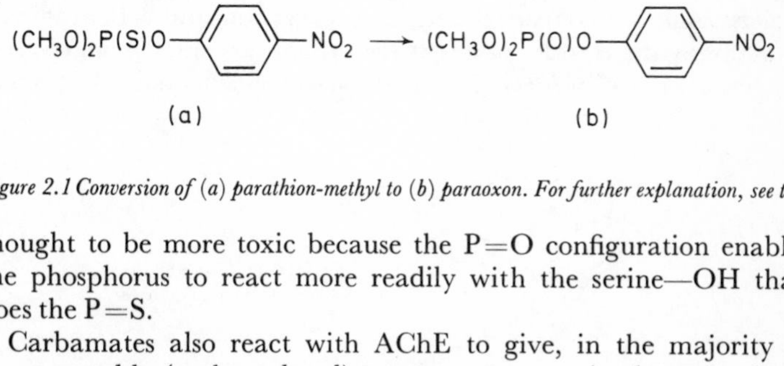

$(CH_3O)_2P(S)O$—⟨benzene ring⟩—NO_2 ⟶ $(CH_3O)_2P(O)O$—⟨benzene ring⟩—NO_2

(a) (b)

Figure 2.1 Conversion of (a) parathion-methyl to (b) paraoxon. For further explanation, see text

thought to be more toxic because the P=O configuration enables the phosphorus to react more readily with the serine—OH than does the P=S.

Carbamates also react with AChE to give, in the majority of cases, a stable (carbamylated) enzyme: once again the serine hydroxyl group is important. Metcalf (1971) has defined carbamate insecticides as 'synthetic analogues' of ACh, and as such considers that they are attracted to the site of action of AChE where they act as substrates for the enzyme. Unfortunately for the victim, the carbamylated enzyme is probably at least a million times more stable than the acetylated enzyme produced briefly during the normal hydrolysis of ACh.

Insecticides may work of course other than by disrupting the transmission of nerve impulses. Rotenone—an active ingredient of derris preparations—is one of these, and worthy of mention here because it is widely used by amateur gardeners in the UK; moreover it is lethal to fish and hence has the potential of significantly affecting vital food chains in the aquatic environment. The compound depresses oxygen consumption in treated organisms and it is now thought to inhibit aerobic oxidation of substrates linked to NAD (nicotinamide adenine dinucleotide). In fact, it inhibits oxidation at the electron transfer step between NAD and flavoprotein, a point of critical importance in the energy cycle of the cell.

When assessing the role of pesticides in pollution, biologists have tended to concentrate on the properties of the toxicants themselves. However, most pesticides are applied in the formulated state, i.e. mixed with other chemicals such as solvents and emulsifiers to enable the active ingredient to be used to the greatest advantage. A number of materials used in formulation are known to have biological activity and hence could be pollutants in the way these have been defined here. In particular, it is of interest to examine the properties of some synergists which, although not markedly insecticidal in their own right, enhance the activity of a number of known insecticides. Sesamex and piperonyl butoxide are widely used for this purpose and are thought to act by inhibiting microsomal oxidases, a group of enzymes responsible for the breakdown of insecticides within organisms. Such synergists contain a methylenedioxyphenyl group and Hennessy (1968) has suggested that this moiety could be converted within an organism into a benzodioxolium ion by the transfer of a hydride to some acceptor in an enzyme system, as shown in *Figure 2.2*. The ion so formed would then act as

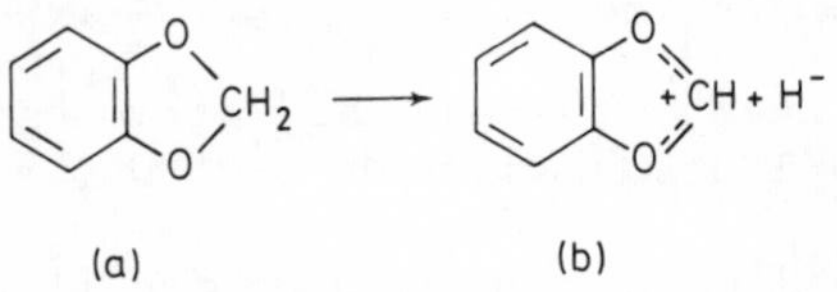

O CH$_2$ O → O +CH + H$^-$ O

(a) (b)

Figure 2.2 Conversion of (a) 1, 3-benzodioxole to (b) the benzodioxolium ion. For further explanation, see text

an acylating agent and would attack, for example, a hydroxyl group (as do organophosphorus insecticides) on a vital enzyme producing a virtually irreversible inhibition. A further interesting feature of the methylenedioxyphenyl synergists is that Bowers (1971) has shown that a number of these compounds are juvenile hormone mimics (see below), but no explanation for this activity is forthcoming as yet.

THIRD GENERATION PESTICIDES: INSECT HORMONES AND THEIR MIMICS

Insect hormones have been chosen as examples of third generation pesticides because their development as pest control agents reflects the philosophy that lies behind this sort of approach, as outlined in the introduction, and because great advances in their exploitation have been made in a relatively short time. Hormones are phsyiologically active chemicals which regulate vital processes such as peristalsis, moulting and water balance. They are produced internally by an organism and transported through the body to target cells which are remote from the site of production. Thus they are different from the neurotransmitters mentioned above, which exert their effects in proximity to where they are produced. They also differ from pheromones, which are secreted externally and effect other *individuals* of the species; these have also found a wide use in pest control in recent years.

The hormone system of special interest in the present context is that concerned with the regulation of post-embryonic growth and metamorphosis in insects. It works in the following way: some factor, which varies from species to species, stimulates certain cells in the brain to produce a hormone (probably a polypeptide) which passes into the insect circulatory system via a pair of glands known as the corpora cardiaca. In the presence of the brain hormone another set of endocrine organs, the prothoracic glands or their equivalents, are stimulated to produce ecdysone which initiates the moulting process in epidermal cells. In immature insects another hormone (juvenile hormone: JH) is present at the moult, which ensures that the next stage in the life cycle will also be immature: it inhibits metamorphosis and hence is absent at the time of the final moult into the adult. Prothoracic glands atrophy in the adult but the corpora allata, which produce JH, become physiologically active again.

Once the chemical nature of ecdysone (a steriod) and JH (a terpenoid) was known, they were synthesised in the laboratory along with their analogues and mimics; indeed mimics of JH were known before the hormone itself was identified. This work established a basis for a novel method of pest control and subsequent research has shown that these materials may at times act as straightforward toxicants (cf. insecticides) and at others fatally upset the moulting/metamorphosis process. However, the present concern is with mode of action rather than toxicity, and it is instructive to examine these hormones from this viewpoint in order to determine their potential as environmental pollutants.

Some of the earliest clues to the way in which ecdysone works came from Beerman and Clever (1964) who observed changes in the morphology of giant chromosomes which are found in various cells of dipterous insects. Following a rise in ecdysone titre, certain of these chromosomes become enlarged (puffed) at specific places along their length and the pattern of puffing changed with time in a predictable manner. Puffs turned out to be sites of messenger ribonucleic acid (mRNA) synthesis, mRNA being of fundamental importance in the synthesis of specific proteins required by cells in their vital processes, as determined by information coded in the arrangement of deoxyribonucleic acid (DNA) of the genes on the chromosomes. This suggested a relationship between a hormone (ecdysone) and the activation or suppression of gene activity: it was the first time for any organism that such a relationship had been demonstrated.

At various stages in the moult cycle, ecdysone can be detected in most tissues of an insect, yet only a limited number of cells (for example in the epidermis) react to its presence: how is this specificity achieved? There is evidence (Gilbert *et al.*, 1971) that the hormone is bound to a specific proteinaceous 'receptor' at the target cell; it also seems possible that α-ecdysone, for example, may be a prohormone and that it is converted into an active metabolite. This has led to the suggestion that a control method could be developed which involved the saturation of protein receptors with inactive molecules: should it prove, as seems possible, a highly specific reaction, it would have a great attraction to environmentalists. More recently it has been discovered that injecting ecdysone can induce a sevenfold increase in the level of cyclic adenosine 3′,5′-monophosphate (*c*AMP). It is known from vertebrate studies that polypeptide hormones stimulate the enzyme adenyl cyclase, in the target cell membrane, which converts ATP to *c*AMP. The latter then acts as a second messenger (the hormone being the first) within the cell, where it elicits a specific response. Adenyl cyclase has been shown to be stimulated in pupal epidermis by β-ecdysone and this is the first example of a steroid producing this effect. Juvenile hormone antagonises the effect of ecdysone on *c*AMP and this is in line with its predicted role as a suppressor of genetic information concerned with metamorphosis.

It is apparent, therefore, that these compounds play a fundamental role in the insect at cell, and even at gene, level. Bearing in mind that many vertebrates have steroid hormones and thus the possibility of target cells sensitive to ecdysone, a great deal of work needs to be done before it can be considered safe to launch such materials on to the general market.

REFERENCES

Albert, A. (1968). *Selective toxicity*, Methuen, London

Beerman, W. and Clever, U. (1964). 'Chromosome puffs', *Scient. Am.* Vol. 210, No. 4, 50–58

Bowers, W. S. (1971). 'Juvenile hormones', in *Naturally occurring insecticides*. Ed. M. Jacobson and D. G. Crosby. Marcel Dekker Inc., New York

Elliott, M. (1971). 'The relationship between the structure and activity of pyrethroids', *Bull. Wld Hlth Org.* Vol. 44, 315–324

Gilbert, L. I., Applebaum, S., Gorell, T. A., Siddall, J. B. and Siew, Y. C. (1971). 'Aspects of research on insect growth hormones', *Bull. Wld Hlth Org.* Vol. 44, 397–398

Hennessy, D. J. (1968). 'Hydride transferring ability of methylene dioxybenzenes as a basis of synergistic activity', *J. agric. Fd Chem.* Vol. 13, 218–220

Holan, G. (1969). 'New halocyclopropane insecticides and the mode of action of DDT', *Nature, Lond.* Vol. 221, 1025–1029

Metcalf, R. (1971). 'Structure-activity relationships for carbamates', *Bull. Wld Hlth Org.* Vol. 44, 43–78

Mullins, L. J. (1955). 'Structure-toxicity in hexachlorocyclohexane isomers', *Science.* Vol. 122, 118

Williams, C. M. (1967). 'Third generation pesticides', *Scient. Am.* Vol. 217, No. 1, 13–17

DISCUSSION

Irvine: Resistance or immunity seems to be building up to some of the older insecticides. Is anything known of the mechanisms of these immunity reactions?

Reay: This is a very complex subject. In some cases there are certainly changes in the enzyme levels within resistant organisms, which can detoxify pesticides. Genetic processes are also involved; an insecticide acts as a selecting agent and, within any population, there will be a number of individuals with a natural ability to handle these materials at what is normally an effective and economic dosage. The small fraction of the pest population which survives will suffer very little intraspecific competition. If the property of resistance is inherited, the next generation will comprise a much larger proportion of resistant individuals and every time the pesticide is applied, this selection will be intensified, tending to increase resistance of the population.

Warrick: We have been using 50 per cent wettable carbaryl to control worms in place of chlordane which we discontinued some while ago on the basis that it is very toxic to birds. Is there very much difference between the two products?

Reay: Faced with a choice, I would use carbaryl. Although it does affect cholinesterase activity it isn't persistent to any extent in the soil so far as I know.

Griffin: Which of the three 'generation' groups of pesticides would you prefer to use if you were engaged in commercial agriculture?

Reay: I should forget all about what is happening at cellular levels and come down to hard economics, which is basically what pest control on food crops is all about. But I wouldn't use a persistent material and apply it every year; I would try to use non-persistent materials where possible. I would use short-lived organophosphates for clean-up, but realise that this is not always possible in the field. The point needs to be made that pollution control is about money—how much are consumers prepared to pay? It should be remembered that pesticides are applied for essentially beneficial purposes: they are not waste products of some manufacturing industry.

Puri: What do you see as the future for pyrethrins?

Reay: Elliot's work on the synthetic pyrethroids has changed the situation radically and it is not without good reason that NRDC has taken out patents on these. They may, in the long run, make us independent to some extent of imported pyrethrum which is very expensive and, in many cases, not so effective.

Warrick: One of the main difficulties with pyrethrins is to store them for any length of time. Do the synthetic pyrethrins have any advantage over the natural product in this respect?

Reay: The natural pyrethrins are not very persistent and although attractive because of this from the pollution point of view, this has restricted their agricultural use. The newer furylmethyl chrysanthemates remain active for longer but this is for days, rather than months or years which is the case with persistent organochlorines.

Mann: Has it been conclusively proven that the organochlorines exert their effect by affecting nerve action?

Reay: No, Holans' work is still very much a theoretical study. From predictions based on his molecular models, he synthesised a number of organochlorine molecules that were DDT analogues and they appeared to have a similar activity. Many of the symptoms of organochlorine poisoning do point to a disruption of axonic transmission but I certainly wouldn't like to give the impression that this is the only type of biochemical lesion they produce.

Knights: What is the future for biological and integrated means of pest control?

Reay: A great deal of difficulty is experienced with these methods in a temperate climate on annual crops because biological control involving, say, a predator–prey relationship, takes some time to become effective, while crop growth times are relatively short. Thus by the time the pest is cleared up, you have lost your crop! Integrated control, apart from the notable exception of some glasshouse applications, is less fashionable nowadays; whether this is because the 'third generation pesticide' approach is more exciting, I don't know. I wouldn't give a value judgment—it is a case of tailoring your control method to particular situations.

Three

The use of herbicides and fungicides

A. Calderbank
Environmental Sciences Group Manager, Plant Protection Ltd., Jealott's Hill Research Station

SUMMARY

Industry, along with government bodies, has a common aim of providing crop protection and pest control methods which are effective and economical, which increase crop yields and safeguard food supplies, and which at the same time give maximum safety to the consumer and to the environment. This chapter outlines the need for control measures, describes the main types of herbicides and fungicides available and the part played by industry in developing new products. The research undertaken to test and evaluate any potential hazards to man, or to his environment, of a new product is discussed with particular reference to the herbicide paraquat and the new systemic fungicide, ethirimol.

NEED FOR CONTROL MEASURES

Pests, weeds and crop diseases have always been with us and man's practice of devoting large areas to a single crop (monoculture) has created very favourable conditions for them to flourish. Weeds and fungi are of major economic importance in competing with man for his food supply; along with insects, nematodes, bacteria and other pests, they cost the human race incalculable damage in direct loss of vital foodstuffs. According to the FAO (Food and Agriculture Organisation), pests destroy up to one-third of the world's food crops during growth, harvesting and storage. Europe is lucky to have a loss in the order of 'only' 25 per cent, whereas in those parts with the greatest food shortage, the proportion is much higher.

The uncontrolled spread of potato blight fungus (*Phytophthora infestans*) in the Irish potato crop during the middle of the last century led to widespread starvation and death, and even today financial losses due to this disease are estimated to be over 20 per cent. Other fungal diseases of major economic importance include the rusts, smuts and bunts of wheat, reckoned to cause annual losses of the order of £60 million in the UK alone. Another fungus *Erysiphe graminis*, responsible for barley mildew, as recently as 1968 cost British farmers 18 per cent of their barley crop, a loss of £35 million worth of grain. Most of the serious diseases of cereals are now well controlled, although rice blast is an ever present threat to the food supply of millions of people in the Far East. Even with control measures, the world losses due to fungal diseases of cereals, fruit, plantation and other crops is estimated at nearly 12 per cent; some 500 million tons of food worth £10 000 million each year (Cramer, 1967).

Weeds compete with crop plants for light, water and essential nutrients. They hamper mechanical harvesting operations and in some cases can harbour harmful insect, virus and bacterial pests. Reduction in crop yield due to weeds can be very heavy, depending on the crop and climatic conditions. Losses of the order of 20–40 per cent are common (FAO, 1967). Herbicides are helping to reduce these losses and are making an important contribution to modern intensive farming operations where labour is scarce and increasing urbanisation is resulting in less land for farming. The intensive and highly productive agriculture of North America, Europe and Japan would hardly be possible without herbicides. As labour becomes scarce and more costly in other parts of the world, the use of herbicides will make an increasingly valuable contribution to agricultural production in the developing countries.

HISTORICAL DEVELOPMENT OF HERBICIDES AND FUNGICIDES

Chemicals have been used for pest control for centuries. The older and often more hazardous products such as those based on arsenic, nicotine, strychnine, rotenone and sulphur, which were widely used in the 1930s, subsequently have been replaced largely by the more effective and often safer organic compounds.

Herbicides

It was during the early days of World War II, when Britain was intensifying her agricultural research in an effort to grow more food, that the possibilities of using selective herbicides with hormone-like activity for controlling broad-leaved weeds in cereal crops was recognised. This resulted in the discovery at Jealott's Hill Research Station of the remarkable properties of MCPA, which along with 2,4-D (see *Figure 3.1*) and related compounds have transformed

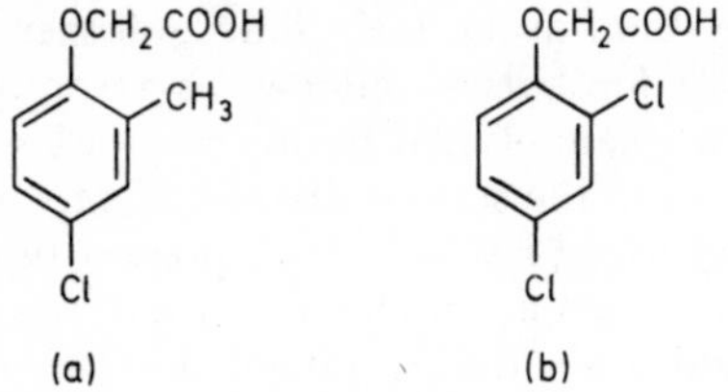

Figure 3.1 Formulae of (a) MCPA and (b) 2, 4-D herbicides selective to broad-leaved plants

cereal growing throughout the world. These are truly selective herbicides in that they can be applied overall to crop and weeds alike and only affect the broad-leaved plants. Cereal crops changed from being one of the weed-prone crops of the farm to being one of the cleanest. The use of these hormone-like weedkillers made it possible to grow cereals year after year on the same ground without a break but, in this situation, grass weeds have greatly increased in importance; for instance wild oat (*Avena* spp) and blackgrass (*Alopecurus myosuroides*). Much research effort is being devoted to solving these problems and already herbicides which will control these weeds are now becoming available.

The important developments in weed control which followed in the 1950s involved the exploitation of chemicals for pre-emergence control of weeds, often in certain economically important crops which showed a specific tolerance to the herbicide. Thus the triazine herbicides, simazine (*Figure 3.2a*) and atrazine, used for weed control in maize, are important examples. The urea herbicides, introduced about the same period are also soil applied. They act mainly on weeds as they germinate and have a period of activity which may last for several months; a notable example is diuron

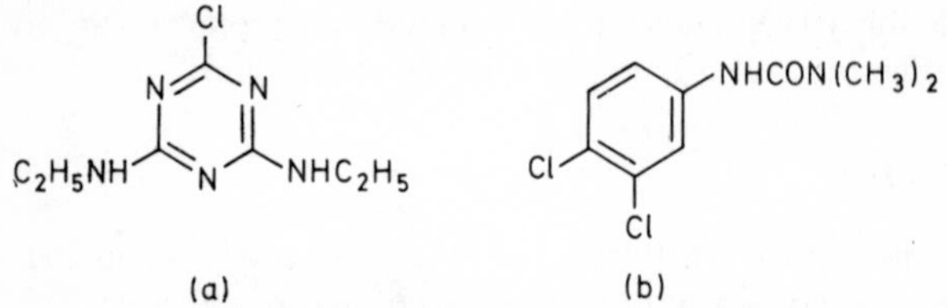

Figure 3.2 Soil herbicides (a) simazine and (b) diuron

(*Figure 3.2b*) which is in wide use today. The range and usefulness of the soil-applied herbicides were extended by the introduction in the 1960s of the dinitroaniline group, of which trifluralin (*Figure 3.3*) is typical. This and others from the same group will control a range of grasses and broad-leaved weeds in a variety of agronomic and horticultural crops. With this armoury of chemicals it is usually possible to find the one which has the greatest margin of safety to the crop whilst controlling most of the infesting weed species.

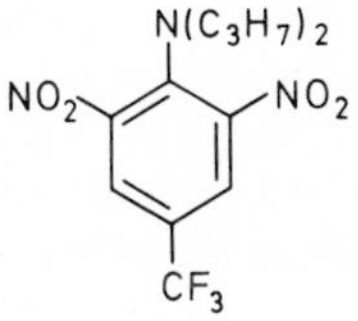

$N(C_3H_7)_2$, NO_2, NO_2, CF_3

Figure 3.3 Trifluralin, typical of the dinitroaniline group of soil-applied herbicides

CH_3^+N ... N^+CH_3 $2Cl^-$

Paraquat dichloride

CH_3^+N ... COOH CH_3NH_2

N-methyl isonicotinic acid   Methylamine

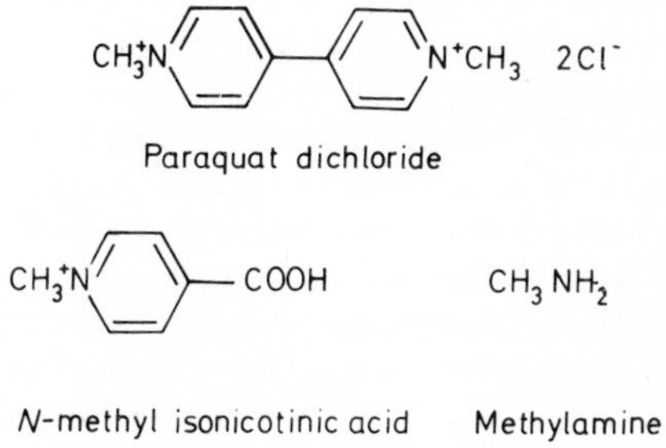

Figure 3.4 Paraquat and its decomposition products

The discovery and development of paraquat (*Figure 3.4*), also in the 1960s, marked a significant technical advance in agricultural practice, not simply by providing more effective means of controlling weeds, but also by offering the possibilities of revolutionary changes in agricultural techniques throughout the world. Paraquat is a contact herbicide which acts rapidly on the foliage and kills both grasses and broad-leaved plants. It is completely inactive in the soil and thus does not affect germinating seed. It can be used to kill established weeds just prior to emergence of the crop plant, or for direct spraying between rows of crops, or in orchards and plantations. Its most significant use, however, is in minimum or non-tillage operations. Weeds or a grass sward can be killed with paraquat and a forage crop such as kale, or a cereal crop, such as corn, or even new grass, can be direct-drilled immediately into the ground with no ploughing or with a minimum of cultivation. The crop emerges free from weed competition. This particular use of paraquat has not only saved labour but introduced much more flexibility into the timing of farming operations. In some situations, notably rice in the Far East, its use has enabled the production of an additional crop during the growing season.

Fungicides

The discovery and development of new organic fungicides to control the important crop diseases has tended to lag behind the rapid progress made in insect and weed control measures and, even as late as 1964, sulphur made up more than 80 per cent of all fungicides used. Fungal diseases of plants are essentially more difficult to control chemically because the fungus is itself a plant living in intimate association with its host. It is consequently difficult to find chemicals which will affect fungi without damaging the crop plant.

Bordeaux mixture (copper sulphate and lime) came into use about 1880 for controlling powdery mildew on vines. Other copper fungicides (for example cuprous oxide or oxychloride) have been developed for use on fruit trees, potatoes, bananas, cocoa, coffee and tea. Inorganic sulphur, because of its low cost, continues to be widely used, mainly on fruit trees, vines and peaches. It is applied as a dust or a spray dispersion of finely divided particles. Lime sulphur, prepared by combining sulphur with lime, is equally effective, and, being soluble in water is easier to apply. It decomposes on the leaf surface leaving a deposit of sulphur. Copper and sulphur fungicides act purely as protectants, however, preventing germinating fungal spores from entering the plant tissue. The crop has to be sprayed repeatedly during the growing season in order to cover new growth and replace the material weathered or leached off the plant by rain.

Mercury compounds came into general use as cereal seed dressings from about 1930. They are very effective, particularly against seed-borne diseases and will also eradicate existing infections as well as protecting crops against further attacks. Their particular virtue is that they will control a wider range of fungal diseases than any other fungicide and at extremely low rates of application. The organic forms of mercury used in agriculture are more toxic than other fungicides, and have recently attracted public attention because of the known accumulation of mercury in the animal body.

The alkyl mercury compounds have, however, been largely replaced by aryl compounds, such as phenyl mercury acetate, which are much safer in use. Much publicity was attached to the US finding that levels of mercury in tinned tuna fish were higher than the permitted tolerance level. It has now been firmly established that the presence of this mercury can be attributed to natural causes (Miller *et al.*, 1972). There are many million tons of mercury present as a natural constituent of sea water and, although the concentration is extremely low, it is capable of being concentrated in certain fish to detectable levels. There is no evidence, however, that these levels constitute a risk to health. Dangerous levels can, however, arise in water and consequently in fish when mercury-containing industrial waste is discharged into restricted areas of water. The quantities of mercury used in agriculture are very much smaller than the waste arising from industrial usage, so the possibility of affecting fish or causing environmental harm from this source can be discounted. In most developed countries the amount of mercury used in dentistry far exceeds that used in agriculture! Despite these considerations and the fact that there is to date no effective replacement for organomercurial fungicidal seed dressings, the use of these materials is being restricted or prohibited in several countries.

A whole range of organic fungicides were developed following World War II. Thiram (tetramethylthiuram disulphide) was one of the first, followed by the dithiocarbamates, for example zineb (zinc ethylene bisdithiocarbamate (see *Figure 3.5*). Thiram, zineb

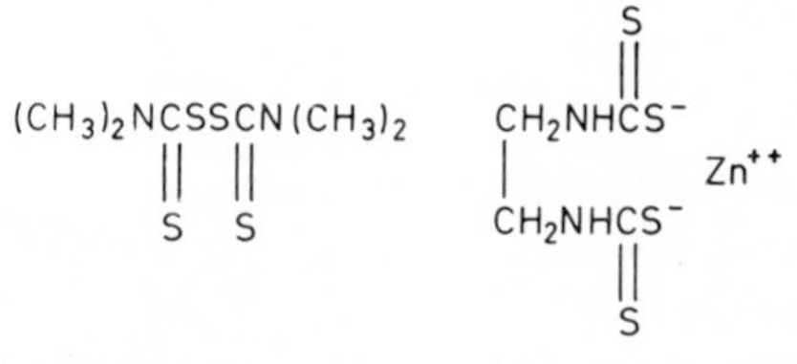

$$(CH_3)_2N\overset{}{\underset{\|}{C}}SS\underset{\|}{C}N(CH_3)_2 \quad (S, S)$$

(a)

$$\begin{array}{l} CH_2NH\overset{S}{\overset{\|}{C}}S^- \\ | \qquad\qquad\qquad Zn^{++} \\ CH_2NH\underset{S}{\underset{\|}{C}}S^- \end{array}$$

(b)

Figure 3.5 (*a*) *thiram and* (*b*) *zineb, widely used as seed protectants*

and other dithiocarbamates are most widely used as seed protectants, but also as foliage sprays for rusts and blights. They are generally of low toxicity and can be used with safety to crops, animals and the environment.

Captan (*Figure 3.6*), developed in the 1950s, has become one of the most versatile fungicides for foliar treatment of fruit and vegetable crops. Several related compounds were also introduced in the

(a) (b)

Figure 3.6 (a) captan and (b) dichlone

succeeding years. They have become widely used to control plant diseases such as scab, rots, mildew, blights, and are very safe from the toxicological and environmental point of view. Other fungicides developed in this period were the quinone group, of which dichlone (*Figure 3.6*) is an example, and the chlorinated nitrobenzenes, for example pentachloronitrobenzene (PCNB). The phytotoxicity of these compounds has been a major difficulty in their development, on a wide scale, as foliar fungicides.

The fungicides described tend to be 'surface acting' and need to be applied frequently during the growing season in order to contain the disease. The past few years, however, have seen the introduction of a new family of systemic fungicides with greater effectiveness and

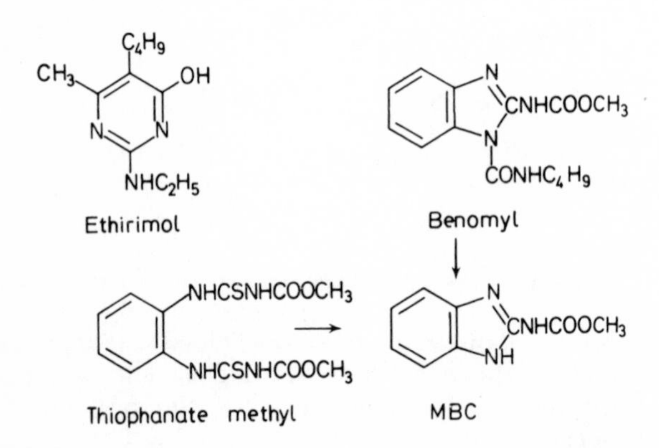

Figure 3.7 Some recently developed persistent systemic fungicides. Benomyl and thiophanate depend for their activity on conversion into the highly fungitoxic MBC

persistence. A typical example is ethirimol (*Figure 3.7*) which controls powdery mildew on barley. The chemical is applied as a seed dressing and gives control of the disease attacking the foliage throughout almost the whole growing season. Benomyl and thiophanate were the first compounds to be introduced with a wider spectrum of fungicidal action, depending for their activity on conversion into the highly fungitoxic methyl benzimidazole carbamate (MBC). Conversion to the active entity occurs on the leaf surface, within the plant and even quite rapidly in the spray solution (Clemons and Sisler, 1969; Selling *et al.*, 1970).

DEVELOPMENT OF A NEW PRODUCT

Many effective and useful new pesticides have been developed and brought into use in the past 25 years to keep pace with new requirements caused, for example, by new pest problems, changing farm practice and the development of resistance to existing compounds. It now takes 6–8 years and costs £2 million or more to develop a new pesticide. Before a new product for agriculture reaches the market today, it must have passed through a whole series of tests, many of which were not even thought of 20 years ago. A great variety of tests for toxicity, residues, breakdown and ecological effects are conducted and there is regular consultation with Regulatory or Health Authorities in the countries where the product is to be introduced. The aim of this work is to ensure that health risks and environmental problems are non-existent or minimal, before a new product is marketed.

From the time when interesting biological activity is discovered in a new chemical and the decision is made to go ahead with development, the chemical is evaluated through a series of tests which will progressively eliminate the compounds which lack the required attributes.

One of the first tasks on the environmental side is to develop a method of analysis for the chemical which should be specific and sufficiently sensitive to measure fractions of less than one part per million of the chemical residues in food crops. The method of analysis is the corner-stone upon which much of the subsequent evidence is built and, therefore, it must not only be highly sensitive, but reliable and reproducible. Masses of figures are useless if they are obtained using a suspect method. Hence the method of analysis itself must be subjected to rigorous scrutiny and tests. The present precision of chemical detection, mainly with the aid of gas–liquid chromatography, exceeds that of the 1930s by three to six orders of magnitude (i.e. a thousandfold to a millionfold increase). Such powerful analytical methods have revealed minute and generally insignificant quantities of some of the more stable pesticides, such as DDT and other materials such as the PCBs, in certain species of wildlife. The significance of these trace residues is largely unknown and the alleged hazards are still hypothetical. It is this uncertainty, combined with the publication of much dubious data, which has resulted in a seemingly exaggerated threat and a real public fear over the impact of pesticides on the environment. Studies must also be carried out on how the product is metabolised in plants or soil, or degraded as a result of external influences, such as exposure to sunlight. Finally, routine analysis for the chemical and its degradation products is conducted on large numbers of crop samples harvested after treatment with the pesticide, both in the development phases and when it is being used commercially, to confirm earlier findings. Residue decay curves must be established on those crops and in those countries where the product is to be introduced so that varying climatic factors and different soil situations can be fully considered.

The work on plants is supported by toxicological studies with various species of animals. Here work is done on the absorption and excretion of the pesticide by animals, and its metabolism within them. This enables information to be obtained about the possible accumulation of the chemical or its metabolites within the animal. The acute toxicity of the chemical to animals is measured, permitting an assessment of potential hazards from the pesticide to operators in the field. An assessment is made of the possible carcinogenicity of the pesticide and of its effects on reproduction. The chronic toxicity is measured by administering the chemical to animals over a prolonged period—frequently as long as two years—and the maximum dose which can be fed without ill-effect is assessed (this is the 'no-effect level'). This level is then compared with typical residues found in samples of crops from field treatments; a large safety margin between the no-effect dose and the crop-residue level is always insisted upon.

Some comment is worth while on the term 'toxicity'. It is not always appreciated that all toxic compounds are not synthetic or 'non-natural'. The natural world is made up of an enormous variety of chemicals, and many of these are, of course, very toxic if consumed in sufficient quantity. Natural materials which are commonly eaten in foodstuffs, such as, for example, caffeine, theobromine and oxalic acid, are as toxic, or even more toxic, than many pesticidal chemicals. For instance, caffeine is much more acutely toxic than most of the herbicides and fungicides in common use today: even vitamin D has been shown to be toxic when concentrations in the diet are high enough. Foodstuffs containing these natural products are not normally considered hazardous because of the low content of the toxicant. Thus all toxicity is relative to dose as is further discussed in a later chapter by Sharratt, and, of course, differences in toxicity to different species are exploited in the development of specific pesticides. D. G. Hessayon, in his recent Tennant Memorial Lecture (Hessayon, 1972), makes this point very forcibly when drawing attention to the natural presence of the known carcinogenic chemical 3,4-benzopyrene in common foodstuffs at levels ranging from 4 to 25 ppm.

The environmental research carried out on the herbicide paraquat and the new systemic fungicide ethirimol, is typical of the type of study which has to be made to establish that a chemical can be used safely, and will not result in harm to man or his environment. An outline of some of the experimental work carried out and conclusions reached will help to illustrate the type of considerations which need to be made in the evaluation of these safety aspects.

Paraquat

One of the most important properties of paraquat is its rapid, almost instantaneous adsorption on to soil particles which means that when it reaches the soil surface it is rendered unavailable to plant roots. This property has been extensively studied and it has been shown that adsorption is predominantly on to the clay particles where it is held by unusually powerful physical forces (Knight and Tomlinson, 1967). Adsorption is so complete that the concentration in solution in the soil falls below the limit of chemical or biological detection. Paraquat cannot be displaced from these strong adsorption sites except by the drastic chemical treatment of boiling the soil with strong sulphuric acid which destroys the clay structure.

This property of being rapidly and completely adsorbed on to soil particles is most important from two aspects:

1. The chemical is rendered unavailable to plant roots and consequently there is no phytotoxicity to growing crops and no residues.
2. In this bound state, it has no effect on other living organisms in the soil and cannot be leached, even with large volumes of water, thus ensuring that it will have no effects on the environment.

Many thousands of crop samples have been analysed for residues of paraquat using a sensitive method capable of detecting as little as 0.01 ppm of the chemical. When used for weed control in orchards and vineyards, for example, or on plantation crops such as bananas, coffee, tea, and when used for pre-crop emergence weed control, inter-row weeding of established crops or minimum-tillage operations, there are no detectable residues in the harvested crops. Thus, when the chemical does not come into contact with the crop during spraying, because of the strong binding to soil, no residues are to be found.

It is only when food crops are treated directly with the chemical, as in pre-harvest desiccation, that significant residues of paraquat are found. Here, the crop plant is treated directly either by spraying the material to be harvested, as with cotton, or a part of the plant that is not harvested, as with potatoes. It is possible with potatoes that the chemical could find its way into the tubers by translocation from the leaves and residues actually measured range from 0.02 to 0.1 ppm. To put this in perspective, it is necessary to compare this figure with the animal 'no-effect' feeding levels. In fact, 110 ppm of paraquat in the total diet has been fed to rats over a two-year period and 36 ppm to dogs without toxic effect. If these levels are compared to the residues which are actually found in potato tubers, and, when account is taken of the fact that potatoes form only a part of the total diet whereas the levels fed to the animals were measured as a proportion of the total diet, there is a thousand-fold safety margin.

The residue studies so far described have given information only about the original chemical applied. It is possible that degradation of the chemical might occur after application with the degradation products entering the soil and possibly subsequent crops. Extensive studies have been carried out using radioactive labelled materials. Paraquat is not metabolised within the plant, but decomposition does occur on the sprayed leaf surfaces due to the action of sunlight. The products of the decomposition have been identified (Slade, 1966) as *N*-methyl isonicotinic acid and methylamine (see *Figure 3.4*, p 29).

Methylamine occurs naturally in plants and consequently attention was focused on the fate of *N*-methyl isonicotinic acid formed on plant leaves by photochemical degradation of paraquat. The toxicity of the compound to mammals was first examined and was found to be quite low (oral LD_{50} to rats > 5 g/kg); following which, long-term feeding tests established that the chemical could be fed at a level of 20 000 ppm in the diet of rats over a 90-day period with no effects whatever. Despite the low toxicity of this compound, attempts have been made to determine its fate in plants and in soil.

Its presence in food crops could be due either to translocation from leaves of plants desiccated with paraquat or to uptake from the soil. Both possibilities have been examined. A sensitive method of analysis was developed and no residues were found in potato tubers harvested from crops desiccated with paraquat, or in wheat grown in soil treated in four successive seasons with paraquat, or in fruit harvested from orchards treated over three successive seasons. The lack of residues of *N*-methyl isonicotinic acid is almost certainly due to its rapid decomposition in soil. Thus, it was shown in laboratory experiments that complete loss of 50 ppm of the chemical occurred in two types of unsterilised soil within 7 days, and in a field plot a drop from 29 ppm to 0.1 ppm over a period of 6 days was recorded.

In addition to the residue and degradation studies in soil and plants as discussed above, other properties of a new herbicide must be investigated to assess possible environmental problems and potential hazards to animals and man in terms of its persistence and fat-solubility. Properties of the chemical, such as its solubility, volatility, chemical stability, general toxicity, excretion characteristics, etc., need to be thoroughly investigated, since all interact and have a bearing on the overall potential hazard.

Paraquat is certainly highly water soluble but is completely insoluble in fat. It is rapidly excreted when ingested by animals and not retained and once in contact with the soil is rendered inert and consequently is not biologically persistent. The degradation products are non-toxic, non-cumulative and are rapidly degraded in the environment, and so there is no possibility of residues being accumulated in living systems. Work at Jealott's Hill has shown that in soils treated at excessively high rates of paraquat (100 kg/ha and above), there were no long-term effects on respiration, organic matter decomposition or general activity of micro-organisms or on the varieties of these organisms normally found in the soil. Effects on micro-arthropods and earthworm populations have been studied and no toxic or repellent effects were observed, even at these excessively high rates of application.

Since paraquat is so tightly bound to soils there is no risk of it being leached by heavy rain or run-off following its use in the terrestial situation. However, the ecological effects of paraquat have been extensively studied in relation to its well-established use in many countries for the control of aquatic weed growth. Paraquat has been shown to disappear quickly from treated water due to its rapid absorption into the weeds which eventually decompose into the bottom mud. It has also been shown to be non-toxic to fish and other aquatic organisms at the low concentrations in the water 0.5–1.0 ppm) needed for weed control (Calderbank, 1970; Way *et al.*, 1971).

The use of any aquatic herbicides raises the possibility of two potential hazards with regard to farm animals: they may drink contaminated water and this may either have a directly harmful toxic effect on the animals or may give rise to residues in the meat or milk and thus indirectly be a hazard to man. Both aspects have been the subject of careful study and experiments have established that sheep and calves can be given up to 20 ppm of paraquat in their drinking water continuously for a month without any signs of ill-effects or effects on normal weight gain. This is many times higher than the initial concentration likely to be present in water treated with paraquat for aquatic weed control.

Stringent tests have also been carried out on calves, sheep and lactating cows to establish whether residues are carried through to the meat and milk. Cattle were shown to suffer no toxic effects over a four-week period when allowed to graze pasture immediately after it had been sprayed with paraquat at a level of 1 kg/ha. The animals continued to graze the herbage which contained residues of 200–400 ppm without any ill-effects and no residues could subsequently be found in the meat or milk. Similar results were obtained by administering large single oral doses of the radioactive labelled herbicides to lactating cows.

Ethirimol

Ethirimol is a systemic fungicide which is particularly active against powdery mildew on barley (*Erysiphe graminis hordei*).It has a relatively low solubility in water (200 ppm) and is translocated upwards in plants by the xylem, but not downwards by the phloem. Physical experiments have also shown that it is adsorbed on to soil particles so that leaching does not occur to any significant extent. To exploit these properties fully, the fungicide is applied to the soil in the root zone, most conveniently as a seed dressing. This forms an area of high concentration in the vicinity of the roots so that small amounts are continually available for uptake by the plant over a period of several weeks. In practice, protection of both old and new foliage can be achieved throughout almost the whole of the growing season.

Extensive residue and metabolism studies have been carried out with ethirimol but because the chemical is applied at a very early stage in the life of the crop, the possibilities of significant residues appearing in the harvested grain, some three months later, are extremely unlikely. This has been confirmed by analysing the cereal grain from trials carried out in England and other part sof Europe. No residues could be found in any of the samples analysed using a method capable of detecting as little as 0.1 ppm of the chemical.

Unlike paraquat, ethirimol undergoes metabolism in plants with a half-life of 3–4 days. A detailed study has also been made (Cavell, Hemmingway and Teal, 1971) of the residues of ethirimol and its metabolites in barley plants and grain following application of ^{14}C-labelled ethirimol as a seed dressing in the field. When plants grown from seed dressed with the radioactive fungicide were analysed, the level of radioactivity in the plant represented 2.2 ppm of ethirimol and metabolites. At harvest the total residue levels in the plant were very low. (This is a consequence of the way in which the fungicide is taken up by the cereal plant from a seed dressing so that the amount of uptake decreases as the plant grows and the material that has already been taken up is progressively diluted.) In the grain a total of 0.04 ppm of metabolites due to ethirimol was present, and of this less than 0.002 ppm was ethirimol itself. In the straw, total metabolites were of the order of 0.12 ppm and of this less than 0.04 ppm was ethirimol.

Ethirimol is a chemical with a low toxicity to mammals, its acute oral LD_{50} to rats being more than 4 g/kg. No adverse effects have been observed in long-term feeding studies with rats and dogs at dietary levels of 1000 ppm. The major metabolites, which arise from *N*-dealkylation, hydroxylation of the butyl group and formation of *O*-glucosides, have been identified and shown to have a similar low toxicity. Metabolism of ethirimol in animals, in fact follows a similar route to that which occurs in plants except that *O*-glucuronides are formed (Daniel, 1971) rather than glucosides. Therefore, there can be no hazard to consumers from residues in cereal grain, since there is more than a thousandfold safety margin between the residue level (including all metabolites) and the highest no-effect feeding level.

The ecological effects of ethirimol in the environment have been studied and there are no effects on soil arthropods or earthworms, and the fungicide seems to be innocuous to soil processes generally. It is non-toxic to honey bees and birds at high rates of application. A survey has been made on fields drilled with cereal seed dressed with ethirimol and no adverse effects have been observed on birds or other animals. A further useful property of the chemical is that it has been shown not to harm predatory ground beetles and spiders which help to control harmful species which attack cereal seed and the young seedlings.

CONCLUSION

The results which have been summarised above illustrate the type of studies carried out on these residue, soil and environmental aspects. They are by no means exhaustive, and much additional experimental work, not described here, has been carried out with ethirimol and with paraquat. It was mentioned earlier that it currently takes from 6 to 8 years from discovery to finally marketing a new product. It is as well to point out, however, that even after all this effort—and as many more years research as could be spent—it is still impossible to say that a product is *completely* safe. All one can say is that a great variety of tests have been carried out which indicate that the compound will have no harmful effects on humans or on the environment and experiences of the past decades, with a great variety of products with diverse properties, have provided us with valuable guides as to the types of experiments needed, so that any remaining risk must be almost negligible and largely hypothetical. It should not need stressing that any slight hypothetical risk from crop protection chemicals must be set against the very real benefits which have resulted from their careful use.

REFERENCES

Calderbank, A. (1970). 'The fate of paraquat in water', *Outl. Agric.* Vol. 6, No. 3, 128–130

Cavell, B. D., Hemingway, R. J. and Teal, G. (1971). 'Some aspects of the metabolism and translocation of the pyrimidine fungicides', *Proc 6th Br. Insectic. Fungic. Conf.* Vol. 2, 431–437

Clemons, G. P. and Sisler, H. D. (1969). 'Formation of a fungitoxic derivative from benlate', *Phytopathology.* Vol. 59, 705–706

Cramer, H. H. (1967). 'Plant protection and world crop production', *Pflanzenschutz-Nachrichten, Bayer.* Vol. 20

Daniel, J. W. (1971). *Industrial Hygiene Res. Labs.* Unpublished results

Food and Agriculture Organisation (1967). Papers presented at the FAO Symposium on Crop Losses. Food and Agriculture Organisation, Rome

Hessayon, D. G. (1972). '*Homo sapiens*—the species the conservationist forgot', *Chemy Ind.* No. 10, 407–411

Knight, B. A. G. and Tomlinson, T. E. (1967). 'The interaction of paraquat with mineral soils', *J. Soil Sci.* Vol. 18, 233–243

Miller, G. E., Grant, P. M., Kishore, R., Steinkruger, F. J., Rowland, F. S. and Guinn, V. P. (1972). 'Mercury concentrations in museum specimens of tuna and swordfish', *Science.* Vol. 175, No. 4026, 1121–1122

Selling, H. A., Vonk, J. W. and Kaars Sijpesteijn, A. (1970). 'Transformation of the systemic fungicide methyl thiophanate into 2-benzimidazole carbamic acid methyl ester', *Chemy Ind.* No. 51, 1625–1627

Slade, P. (1966). 'The fate of paraquat applied to plants', *Weed Res.* Vol. 6, 158–167

Way, J. M., Newman, J. F., Moore, N. W. and Knaggs, F. W. (1971). 'Some ecological effects of the use of paraquat for the control of weeds in small lakes', *J. appl. Ecol.* Vol. 8, 509–532

Four

The effects of pesticides on wildlife *

Abstract of a paper presented by
F. Moriarty
Monks Wood Experimental Station

Organochlorine pesticides are good examples of persistent pollutants. Despite the relatively small amounts present in animal tissues, skilled analysts can make quite accurate determinations. Differences between samples are a much more important source of variation.

The residues found in animals after exposure are consistent with the compartmental model. Aquatic species acquire their residues by direct uptake from the water, and that acquired with food appears to be of minor significance. In other words, position in the food chain or food web is irrelevant.

The situation is less clear-cut for terrestrial species, but here too the main emphasis should be placed on the species' rates of uptake and loss.

DISCUSSION

Warrick: With organochlorine insecticides, is the rate of absorption through the skin very high in terrestrial species?

Moriarty: It depends very much on the formulation. For example, rate of entry from a crystai of insecticide placed on the cuticle of an insect is very slow compared with the same amount of material applied in a suitable solvent.

Newman: *Plant Protection Ltd.* Am I right in thinking that in

* This talk was based on a fuller text entitled 'The effects of pesticides on wildlife: exposure and residues' published in *The Science of the Total Environment*, Vol. 1, 267–288 (1972)

your view there is severe doubt about food chain accumulation in aquatic animals but that you have an open mind about the terrestrial situation? Have you any views on the Swedish work on the accumulation of methyl mercury in aquatic systems from paper mill wastes?

Moriarty: I do not think there is any good evidence to show that an aquatic organism's position in the food chain affects directly the amount of organochlorine insecticide that it contains.

For terrestrial predators, we really don't know, but I prefer the emphasis to be placed on the individual species' physiology: its rate constants for uptake and elimination of these compounds, and the effects of other factors on these rates. I think we will then be able to develop a more useful approach for predicting whether or not a particular compound is likely to cause problems.

To answer your second question, Hannerz published a review in 1968 (Report No. 48, Institute of Freshwater Research, Drottingholm, 120) in which he concluded that there was no evidence for accumulation of mercury up the food chain in aquatic systems. It is true that Jernelöv and Lann (*Oikos*, Vol. 22, 403–406 (1971)) deduced that the amount of methyl mercury in fresh water species depends partly on that which they acquire with their food, and partly on that which they acquire directly from the water. For example, 60 per cent of the methyl mercury found in pike (*Esox lucius*) comes from their food. However, this calculation involves at least two assumptions which may be invalid. Predatory fish in laboratory experiments accumulated 10–15 per cent of the mercury that was in their food. Jernelöv and Lann assumed this figure is true for field situations. They also assumed that the intakes from food and from water are additive, i.e. that the one source does not affect the intake from the other source. Chadwick and Brocksen's (1969) data (*J. Wildl. Manag.*, Vol. 33, 693) for dieldrin suggested that there is an interaction.

Five

The run-off of fertilisers from agricultural land and their effects on the natural environment

E. R. Armitage
Crop Production Unit, Agricultural Division, I.C.I. Ltd.,
Jealott's Hill Research Station

SUMMARY

Fears have been expressed about possible pollution of natural waters and reservoirs by high concentrations of nitrogen, potassium and phosphate nutrients resulting from the run-off of modern chemical fertilisers from agricultural land. Potassium has limited mobility in soils, however, whilst phosphorus is virtually immobile and therefore neither are leached-out very easily and do not appear to have any serious effects on the natural environment or on human health.

Nitrogen fertilisers on the other hand are readily converted into nitrates which are soluble and these pose more serious problems. Studies show, however, that a proportion of the nitrates are attributable to the activity of nitrogen-fixing bacteria found in all fertile soils whilst some is present in rain water and in sewage effluents. Evidence reviewed indicates that the level of nitrate nitrogen in some rivers in the UK is rising but need not at present give cause for serious concern. Certain river and well waters which occasionally have very high levels do require careful monitoring. In addition, problems also arise in the disposal of large amounts of manure and slurries from the increasing numbers of intensive animal rearing units.

Although pollution of water by fertilisers is not a serious problem at the present time, a number of possible means of meeting problems in the future are outlined.

INTRODUCTION

In recent years there has been understandable concern about environmental pollution and it has been frequently stated that modern fertiliser usage has added to this pollution by increasing the concentration of fertiliser nutrients in drains, rivers, lakes and reservoirs. This chapter deals in some detail with the possible losses of fertilisers into water-ways and in less detail with the effect on the environment, and outlines briefly ways of overcoming any serious problems which might arise. The term 'fertiliser' refers to both the chemicals manufactured in large tonnages and also the animal manures, since they are taken up by the plant in a more or less identical manner.

The tonnages of manufactured fertiliser nutrients, nitrogen, phosphorus and potassium, used in the UK in recent years are given in *Figure 5.1*. These data, which are readily obtainable from the

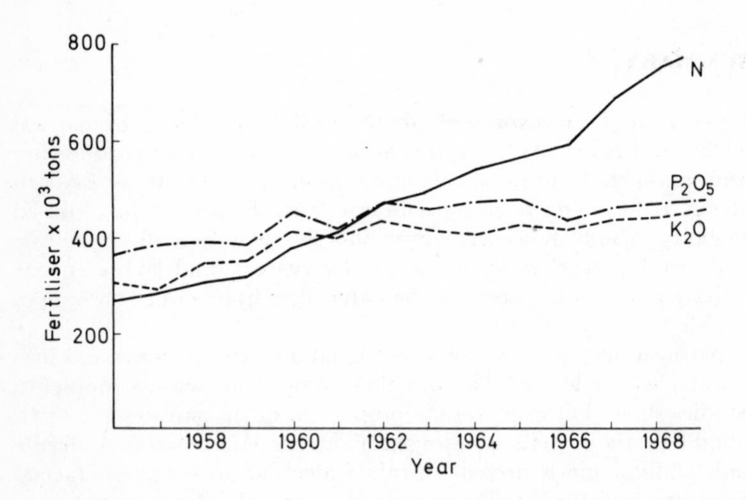

Figure 5.1 Fertiliser usage in the UK 1958–69

Ministry of Agriculture, Fisheries and Food returns, show that the usage of both phosphorus and potassium has remained fairly steady at about 400 000 tons (approx. 406,000 tonnes) (as P_2O_5 and K_2O) per annum while the nitrogen usage has almost doubled in the last 14 years and is now about 800 000 tons (approx. 813 000 tonnes). The tonnages of nutrients from farmyard manure are more difficult to obtain but estimates for 1968 give similar weights of nitrogen and phosphorus, but the levels of potassium are about three times those in conventional fertilisers.

SOURCES OF FERTILISER NUTRIENT LOSS

Fertiliser nutrients can be lost from farmland in three different ways: first, by drainage water percolating through soil, leaching soluble plant nutrients, secondly, by the inefficient return to the land of the excreta of stock and, thirdly, by the erosion of surface soils or the movement of fine soil particles into subsoil drainage systems. Before it is possible to assess the losses of these nutrients, their action in the soils, especially where they behave differently, has to be considered.

Nitrogen

No matter in which form the nitrogen is applied, and this includes farmyard manure, it is taken up by the plant as water soluble ammonium or nitrate ions which are produced by microbial action on the organic form. The ammonium nitrogen is also readily converted microbially to nitrate and it is this which will be discussed in most detail. Nitrate nitrogen is not held by the soil particles since they are mainly negatively charged and any nitrate which is not taken up by plant roots is readily leached into the subsoil and from there into the drains, rivers or underground aquifers.

Phosphorus

The form in which the phosphorus is applied in conventional fertilisers, which is taken up by the growing plant, is as the orthophosphate. On addition to the soil, phosphates are readily precipitated as either calcium, iron or aluminium phosphate depending on the nature of the soil. The phosphate is then slowly released from these insoluble compounds or by the decomposition of organic matter. As a result, phosphate is not mobile in the soil.

Potassium

Potassium behaves differently in the soil. The potassium ion may sometimes be incorporated from solution into the lattice of some clay minerals, but more usually its availability for plant uptake, and also for movement through the soil, is not so much influenced by the formation of insoluble potassium compounds as by the electrostatic attraction to the negatively charged clay colloids. This process is known as base exchange. The exchangeable potassium can be taken up by the plant and while it can be replaced by other cations such as calcium, its mobility is limited and it tends to become re-adsorbed onto the clay colloids.

Thus the three nutrients applied to the soil have very different actions. Nitrogen as nitrate is not adsorbed and is very mobile, potassium is readily adsorbed and has limited mobility, while phosphorus is precipitated and is virtually immobile.

THE MEASUREMENT OF LEACHING LOSSES

There are two main methods of measuring leaching losses and both have certain limitations. First, concentrations of the fertiliser elements in field drains can be measured. By this method, provided it is possible to obtain a suitable method of sampling, an accurate measurement of the concentration in the solution can be obtained. However, because it is impossible to measure the efficiency of the drains and to measure accurately the area from which they are extracting water, it is only possible to obtain an estimate of losses on an area basis.

The second method is by the use of lysimeters. These are encased blocks of soil, sunk flush into the ground, from which the drainage water can be collected. They have the disadvantage that they are limited in size and there is the risk of uneven drainage down the sides of the casing. Also the rate of drainage may differ from natural conditions because of the loss of capillary rise. Classical examples of lysimeters in the UK include those at Craibstone, near Aberdeen, and at Rothamsted. The latter were set up in 1870 and have since been maintained in a fallow weed-free condition and no fertilisers have been applied. In more recent times, lysimeters have been in operation at The Grassland Research Institute, Hurley, at Great House Experimental Husbandry Farm and at Jealott's Hill Research Station.

Potassium

While potassium has been studied less than either nitrogen or phosphorus, the leaching of potassium was measured in the early experiments at Rothamsted and in 1882 it was concluded that potash was very perfectly retained, the part unassimilated by the crop being held in the soil, chiefly in its upper layers. The annual losses obtained were between 3 and 9 kg K/ha (Lawes, Gilbert and Warington, 1882). The soil at Rothamsted is heavy and the mechanism of retention of potassium by ion exchange on the clay mineral surfaces suggests that leaching losses may well depend upon the nature of the soil. Thus the Craibstone lysimeters, which have a coarse sandy soil of low clay content, lost some 11 kg K/ha/annum (Hendrick and Welsh, 1938).

Most of the evidence available, however, indicates that the annual potassium losses in the drainage water are below 10 kg K/ha. The average losses from the Great House lysimeters were under 3 kg K/ha, even where 125 kg K/ha has been applied. Similarly at Jealott's Hill, the losses were even smaller averaging some 2 kg K/ha.

Although many factors influence the possible movement of potassium in soil, such as cultivations, liming, the application of elements other than potassium and the placement, nature and, in particular, the rate of fertiliser application, it would appear that at present the danger of large losses of potassium in the drainage is slight. Studies on this subject have been somewhat neglected in the past and it would be of interest to find out more about the concentration of potassium in rivers and to ascertain if they are changing with time.

Phosphorus

As mentioned earlier, phosphates are readily precipitated and are virtually immobile in the soil. If the soil is extremely sandy, however, it is possible that there may be insufficient bases to fix phosphorus. Such soils would have a low total phosphorus content and some applied fertiliser phosphate could be lost in the drainage. There is only one reported case in this country and this occurred in Wareham, Dorset, on very sandy soil containing practically no clay (Gasser and Bloomfield, 1955).

The figures reported from both lysimetry and drainage studies all give very low levels of phosphorus in the drainage water. No phosphorus was found in the leachate from the Craibstone lysimeters in the six-year period, 1921–26, while at Jealott's Hill the losses from grass-cropped lysimeters were under 1 kg P/ha/annum.

The concentration of phosphorus in field drains at Saxmundham, on the chalky boulder clay in Suffolk, and in the drains at Woburn averaged about 0.1 ppm or less over a year. All this information indicates that the losses of phosphate by drainage are very small.

Nitrogen

Nitrogen is more likely to be leached than either phosphorus or potassium because of the mobility of the nitrate ion. Most agricultural soils contain between 0.075 per cent and 0.3 per cent of total nitrogen, in other words they contain between 1500 kg and 6000 kg N/ha in the top 150 mm. Almost all this nitrogen is combined with about 10 times as much carbon in soil organic matter. Light sandy soils and soils under continuous arable cultivation contain least nitrogen, more is found in heavier clay soils under permanent grassland. Poorly drained soils also tend to accumulate organic matter and the fen soils in Eastern counties can contain as much as 1 per cent total nitrogen.

In all soils a small proportion of the nitrogen associated with the organic matter is mineralised annually to the ammonium ion which is then quickly oxidised to nitrate. Normally, in old arable soils, between 30 and 60 kg N/ha may be released each year, but where land receives large amounts of organic matter or crop residues, and in pastures containing legumes, much more nitrogen may be released. The release is due to microbial action and the rate of release is dependent on many factors and is greatest when the soil is moist and warm. A flush usually occurs in the spring and there may be another flush in the autumn when dry soils are rewetted or fresh crop residues are added.

These supplies of available nitrogen are almost invariably insufficient for optimum yield and are usually supplemented by the addition of fertilisers containing ammonium and/or nitrate nitrogen. The level of nitrogen usage is increasing and at present averages about 75 kg N/ha. The actual amounts applied will vary with cropping and intensity of farming and up to 350–400 kg N/ha can be applied to very intensive grassland. However, on well farmed land the amount of nitrate from both fertiliser and from the mineralisation of soil reserves amount, on average, to about 200 kg N/ha.

It must be emphasised that the nitrate from fertilisers and that from soil reserves is indistinguishable and as there is no mechanism to retain nitrate in soils, all the nitrate is at risk. That which is not taken up by the growing plant and its root system will be lost. The loss can occur either as nitrate in the leachate or by microbiological denitrification in a gaseous form. A very approximate estimation of nitrogen loss can be obtained from the nitrogen recovery, i.e. the increase in nitrogen uptake in the treated crop over that in the control. This method makes no allowance for any possible extra nitrogen uptake in the fertiliser/root system. Arable crops usually recover no more than 50 per cent of the fertiliser nitrogen while grass may recover up to 80 per cent, but the amount is very variable.

Nitrogen losses—lysimeter studies

The various lysimeter experiments have provided a considerable amount of information about nitrate losses in the drainage water. The oldest lysimeters at Rothamsted have been maintained in a fallow condition for over 100 years and are still losing nitrate nitrogen although no fertilisers have been applied. The only nitrogen which they have received in this period has been a small amount in the rainfall, averaging about 5 kg N/ha per annum. Averaged over the period 1878–1905, the drainage water contained about 10 ppm nitrate nitrogen, equivalent to an annual loss of 35 kg N/ha (Miller, 1906). Even after 50 years, loss was some 23 kg N/ha/annum and the level of total nitrogen in the soil had dropped by one-third from about 0.15 per cent to 0.10 per cent (Russell and Richards, 1920). The size of the nitrogen reserve in the soil can be realised when after 100 years the average level of nitrate nitrogen in the water was approximately 5 ppm, about half the original value and the annual rate of loss was almost 20 kg N/ha.

In the Craibstone lysimeters the average losses of nitrogen for a six-year rotation were about 8 kg N/ha and there were virtually no differences in losses between those lysimeters receiving no fertiliser nitrogen and those receiving some 224 kg nitrogen during the rotation. Also the annual losses of nitrogen for the years under pasture management were lower and averaged about 2.5 kg N/ha /annum. Although the Craibstone lysimeters have provided some valuable information on a coarse sandy soil, it is somewhat difficult to make a detailed comparison with modern conditions because, by present day standards, the levels of fertiliser applied, particularly those of nitrogen, were very low.

At Great House Experimental Farm on the Lancashire slopes of the Pennines, drainage losses are being measured on a poorly drained peat soil under turf. In the period 1963–67 virtually no nitrogen was lost from lysimeters receiving no fertiliser nitrogen, but about 25 per cent was lost from those lysimeters receiving some 224 kg N/ha/annum (Anon, 1970).

At Jealott's Hill, replicated monolith lysimeters were used to examine the nitrogen losses under different croppings in the period 1951–56. During this period, no fertiliser nitrogen was applied and

Table 5.1 RESULTS OF ANALYSES OF WATER DRAINAGE AND NITROGEN LOSSES FROM LYSIMETERS AT JEALOTT'S HILL, SEPTEMBER 1951–JANUARY 1956 (*After Low and Armitage, 1970*)

Lysimeter treatment	*Total drainage* (mm water)	*Water drainage as % of rainfall*	*Nitrogen losses* (kg N/ha)	*Concentration of nitrogen in drainage water* (ppm N)
Fallow	1397	52	504.4	36.1
Clover	828	31	193.9	23.5
Grass	752	28	12.5	1.7

the only addition was from the rainwater. There were three different croppings, grass, clover and a fallow and the mean total volumes and levels of nitrogen in the drainage for the $4\frac{1}{4}$ year period are given in *Table 5.1*.

A feature of the results was the very low level of nitrogen lost from the grass cover averaging about 3 kg N/ha/annum, while the fallow plots lost a very high level of 120 kg N/ha/annum which was considerably more than the appreciable amount, 46 kg N/ha/annum, lost from the clover cover. The concentration of the nitrate nitrogen in the leachate from the fallow lysimeter averaged some 36 ppm.

Nitrogen losses—field drain studies

Workers at Rothamsted, and its associated outstations at Woburn and Saxmundham, have studied the nitrogen losses in field drains for many years. In the famous Broadbalk experiment, wheat has been grown continuously since 1843 and the composition of water in the tile drains of the fields has been measured. Plots which have received recently about 95 kg/ha of fertiliser nitrogen have been found to lose about 17 kg/ha in the drainage with the average concentration at about 7 ppm. Even when no fertiliser was applied there was still an annual loss of 11 kg N/ha (Cooke and Williams, 1970). Similar results have been obtained from the all-arable site at Saxmundham where, after occasional periods of very heavy rain in the spring, the nitrate concentration in the drainage reached peak levels as high as 50 ppm N.

More recent data on the effect of management on the nitrate concentration in drainage water from Woburn are given in *Table 5.2*.

Table 5.2 RESULTS OF ANALYSES OF WATER DRAINAGE AND NITROGEN LOSSES FROM LAND DRAINS AT WOBURN, MARCH 1968–MARCH 1969 (*After Cooke and Williams, 1970*)

Land drain from	*Nitrate* N *in drainage water* (ppm)		*Phosphate* P *in drainage water* (ppm)	
	Average	*Range*	*Average*	*Range*
Intensively farmed sandy soil, arable-grass	22.5	16–26	0.08	0–0.30
Sandy soil, arable	12.2	10–16	0.03	0–0.51
Grass on drift over Oxford clay	8.0	1–24	0.08	0–0.75
Rough grassland	3.3	1–10	0.08	0–0.30

The level of nitrate from land under grass was again lower than from under arable conditions, and the low phosphorus level in the drainage is also recorded.

Nitrate nitrogen losses generally have been found to be considerably higher under fallow than under plant cover, even when no nitrogen fertiliser has been applied. This might be expected since the nitrate ion is very mobile and if there were no uptake by the plant, it would be leached readily. Arable cropping is intermediate between fallow and continuous plant cover, but in the autumn and winter period, when the volume of drainage is greatest, the plant cover will normally be at a minimum. So nitrate losses under arable conditions will be probably nearer fallow than complete cover. Although the levels of nitrate nitrogen loss under all-grass cover appear to be low, there have been few experiments where they have been studied under the heavier applications of nitrogen which are recommended for maximum production. At Jealott's Hill a long-term study has been started of annual nitrogen losses in drainage from both lysimeters and field drains using annual nitrogen applications up to 600 kg N/ha.

THE EFFECT OF ANIMAL EXCRETA AND EROSION

Losses of phosphate and potassium by percolation through the soil are relatively slight and dependent on management, cropping and fertiliser treatment but nitrate losses can vary from slight to considerable. Besides these percolation losses, nutrients can be lost from farmland both by the inefficient return of the excreta of stock and by the surface erosion of soils. All nutrient elements will be at equal risk from erosion losses.

Farm animals contribute in their excreta considerable quantities of plant nutrients. Normally, with extensive grazing, any nutrient loss will be by percolation and should be similar to that from corresponding fertiliser dressings. With increases in intensive rearing (so-called 'factory farming'), which in particular affects pigs and poultry and to a lesser extent winter housed cattle, the disposal of excreta is a growing problem. With decreasing availability of farm labour, manure is expensive to prepare and handle. Much of this animal waste is converted into semi-liquid slurries which can be returned to the land and if this is done carefully and spread thinly and evenly over the land, no more nutrients should appear in the drainage than from the equivalent amount of fertilisers. If applied in winter time, however, particularly on frozen ground or before heavy rain, there is a serious risk that a considerable amount of the nutrients could find their way into water courses by surface run-off. This problem is well recognised and the disposal of farm excreta is a major factor in present-day animal management. Despite the risk it is considered at present that more of the nitrates found in rivers have originated from leaching and soil drainage than from run-off of animal waste.

The loss of nutrients by wind and water erosion is very difficult to measure and no work has been done in this country, where it is, as yet, not a major problem. Most streams, however, carry considerable quantities of suspended soil material especially after heavy winter rains. This is a source of fertiliser nutrient loss which should not be overlooked. Also, fertiliser nutrients find their way into watercourses from sources other than agricultural ones and sewage effluents in particular may make substantial contributions to both nitrate and phosphate levels.

Owens (1970) studied the nitrogen and phosphorus concentrations in sewage effluents flowing into an 80 kilometre stretch of the Great Ouse from the headwaters to Tempsford, below Bedford. He calculated that only about 30 per cent of the nitrogen in the river came from sewage outflows, which had an average concentration of about 30 ppm nitrate nitrogen. He also found that the bulk of the phosphorus, about 95 per cent in the river water, came from the sewage.

THE EFFECTS OF HIGH NUTRIENT LEVELS IN WATER

Although the fertiliser loss is important economically to the farmer, the more important concern is its effect on the environment.

There are two major problems which have been associated with high nutrient levels. First, algal blooms have been attributed to soluble nitrates and phosphates in water. The causes of such growths, which have been known for centuries, are incompletely understood but the levels of nutrients which it is claimed limit algal growth are below 0.3 ppm N and 0.01 ppm P respectively. In most lowland waters in the UK nutrient concentrations have been above these limits for decades but blooms have been rare and short lived. It is thought that factors such as water temperature, carbon dioxide concentrations and the presence of organic matter are important and, in general, it seems unlikely that marginal increase in nutrient levels would initiate algal growth.

Secondly, increased nitrate levels can cause a health hazard in drinking water for very young babies up to the age of about three months. The nitrate in the water can be reduced in the baby's stomach to nitrite which combines with the haemoglobin to form methaemoglobin which prevents the transport of oxygen around the body. This is one of the causes of 'blue babies' and about 10 cases of methaemoglobinaemia have been confirmed since the condition was recognised 20 years ago. These have all been associated with

8, indicate the most common reason.
le (in circulation, marked missing, or not on

ion

nation you find in the library?
Inadequate [ ] Undecided

formation you find in the library?
opriate level [ ] Too easy or too popular

ours to meet your needs?

stance?
5)

taff helpful?
question 16)

aff is too busy [ ] I can't ever find the library staff

___________________.

ease check the appropriate box:
adequate (I), Undecided (U)

?) (E) (I) (U)

ground waters, usually with water from shallow wells, which may have been contaminated with sewage. Recently, there may have been two cases arising from a public supply in Lincolnshire (from wells in the chalk) and concern has been expressed about the nitrate levels in undergound well-water obtained from the chalk areas of the Yorkshire Wolds. This aspect will have to be kept under careful scrutiny in the future. The World Health Organisation has therefore recommended drinking water standards for Europe and these are given in *Table 5.3.*

Table 5.3 WORLD HEALTH ORGANISATION: RECOMMENDED DRINKING WATER STANDARDS FOR EUROPE (*Source: 'European Standards for Drinking Water', WHO, Geneva, 1961*)

Concentration (ppm)		*Standard of water*
Nitrate	*Nitrate* N	
0– 50	0–11.3	Recommended
50–100	11.3–22.6	Acceptable
above 100	above 22.6	Not recommended

LEVELS OF NITRATE NITROGEN IN RIVER WATER

Because of the importance of the potential environmental pollution problem and in order to ascertain if levels of nitrate were rising, Tomlinson (1970), with the active co-operation of various Water Boards, carried out a comprehensive survey of the nitrate contents of some 17 rivers in England, over the period 1953–67. In these areas, water authorities obtained drinking water from the rivers which are frequently monitored and a reasonable assessment of the nitrate content was obtained. In addition, Tomlinson attempted to investigate any possible correlations between nitrate levels in the water and fertiliser usage in the approximate catchment areas.

There are obvious difficulties in the presentation of the mass of figures and a derived mean annual nitrate concentration was used to ascertain changes with time and these were subjected to statistical interpretation. The results are given in *Table 5.4.*

In 7 of the 18 comparisons the nitrate concentration increased significantly with time and although coefficients were positive in 7 other cases there can be little confidence in these correlations. In 4 cases the correlations were negative although only one, the Dee at Chester, had any statistical significance.

The mean annual levels were relatively low and varied from 6.2 ppm for the Stour at Langham, in Essex, to as low as 0.4 ppm for the Tyne. As averages, they did not show peak levels when

Table 5.4 RATE OF INCREASE OF NITRATE CONCENTRATION IN ENGLISH RIVERS BETWEEN 1953 AND 1967 (*After Tomlinson, 1970*)

River	*Mean concentration of nitrate* N *in 1967* (ppm)	*Rate of increase between 1953 and 1967* (ppm *per annum* ±90% *confidence limits*)
Rother (Sussex)	3.1	+0.31±0.09†
Severn	3.1	+0.09±0.03‡
Stour	6.2	+0.15±0.06
Dove	2.8	+0.07±0.04
Blithe	2.7	+0.06±0.05
Wensun	4.3	+0.09±0.07
Devon	5.3	+0.14±0.13
Poulter	4.1	+0.10±0.10
Kennet	2.6	+0.03±0.04
Dove (2nd sampling point)	2.3	+0.05±0.07§
Derwent	0.9	+0.02±0.03
Manifold	1.5	+0.02±0.04†
Thames	5.0*	+0.02±0.05
Tyne	0.4	+0.01±0.03
Leen	3.4	−0.04±0.07‡
Great Ouse	4.2	−0.04±0.08
Tees	0.5	−0.03±0.05 ‖
Dee	1.1	−0.06±0.04§

* Concentration in 1964
† 1956–67
‡ 1957–67
§ 1961–67
‖ 1958–67

Table 5.5 THE RANGE OF MAXIMUM LEVELS OF NITRATE N IN ENGLISH RIVERS OVER A NUMBER OF YEARS (*After Tomlinson, 1970*)

River	*Sampling point*	*Period*	*Range of maximum concentrations* (ppm)	*Correlation coefficient of increase in concentration with time*
Rother*	Horsham	1961–67	1.7– 3.4	+0.83†
Severn	Tewkesbury	1957–67	3.2– 7.7	+0.69†
Stour	Langham	1953–67	7.8–21.8	+0.60†
Dee	Chester	1961–67	2.5– 4.5	+0.25
Great Ouse	Bedford	1953–67	4.8–16.4	+0.05
Tyne	Wylam	1956–67	0.6– 2.0	−0.30
Tees	Darlington	1958–67	0.8– 2.5	−0.49

* Highest monthly mean value
† Correlation significant at the 10 per cent level

nitrate contents could have been much higher. A more detailed study of the maximum levels of 7 of these rivers is given in *Table 5.5.*

As expected, the maximum concentrations were considerably higher than the annual means, especially for the river Stour, where maximum levels recorded in 1965 and 1967 exceeded 20 ppm nitrate nitrogen. There was also a significant increase with time in the maximum levels in the Stour, the Severn and the Rother, which is a tributary of the Arun in Sussex. Tomlinson's survey also indicated, as was claimed by Owens in his work on the Great Ouse, that agricultural land contributed more nitrates to rivers than sewage. This is illustrated in *Figure 5.2* which gives the mean

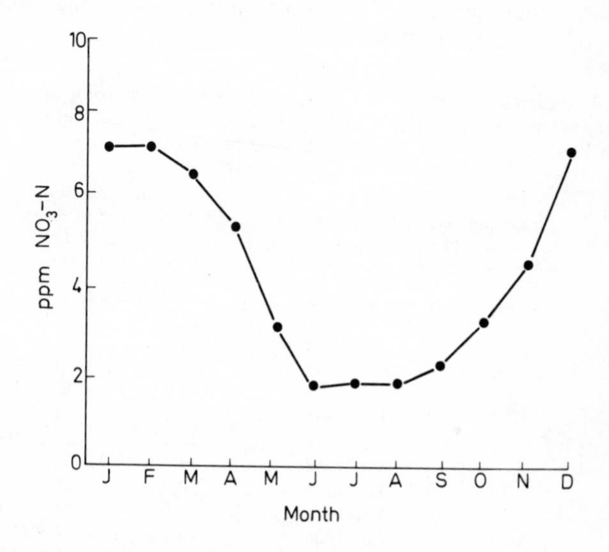

Figure 5.2 Average monthly nitrate concentrations (ppm *nitrate* N) *in the River Stour at Langham between 1957 and 1968* (*After Tomlinson, 1970*)

monthly nitrate concentration in the river Stour. The nitrate concentrations were greatest in the winter months and lowest in the summer. If sewage had made a major contribution to the nitrate concentrations it would have been expected that, since the level in the sewage is constant throughout the year, the nitrate concentration in the winter would be lowest because of the increased volume of flow.

With the increasing use of fertiliser nitrogen it is important to consider how far the increased usage contributed *directly* to the tendency of river nitrate levels to rise. Measurement of the correlation coefficients between nitrate levels and the fertiliser consumption in the catchment areas of rivers was not helpful because both factors are correlated with time and would therefore be expected to be correlated with each other, as indeed they were, without there necessarily being any causal connection between them.

It was considered more meaningful to compare the actual amount of nitrogen fertiliser applied to the catchment area with the quantities of nitrate carried by the river and the distribution of both throughout the year. This was done for the river Stour, which had given the highest nitrate figures of all rivers studied. The total weights of nitrate nitrogen carried by the river, over a 12 year period from the catchment area of 223 square miles, are tabulated in *Table 5.6* together with the fertiliser purchases for the adjacent counties of Essex and Suffolk.

Table 5.6 QUANTITIES OF NITRATE N CARRIED BY THE RIVER STOUR IN RELATION TO THE QUANTITIES OF FERTILISER USED IN THE RIVER'S CATCHMENT AREA BETWEEN 1956 AND 1968; SEE TEXT FOR FURTHER EXPLANATION (*After Tomlinson, 1970*)

	Quantities of nitrate N *in river* (kg N/ha *of catchment area, 223 square miles*)		*Fertiliser purchases* (kg N/ha *of catchment area*)	
Year	*July–June*	*March–May*	*Essex*	*Suffolk*
1956–57		0.6 (1957)	32.6	33.6 (1957)
1957–58	6.5	1.4 (1958)	35.2	31.8 (1958)
1958–59	5.6	0.7 (1959)	34.9	33.4 (1959)
1959–60	8.1	1.6 (1960)	43.9	42.5 (1960)
1960–61	10.2	1.0 (1961)	42.1	44.7 (1961)
1961–62	7.2	1.0 (1962)	45.4	45.2 (1962)
1962–63	5.3	3.8 (1963)	44.7	45.2 (1963)
1963–64	7.3	4.4 (1964)	50.0	53.0 (1964)
1964–65	3.1	1.8 (1965)	51.5	54.5 (1965)
1965–66	10.3	2.2 (1966)	49.4	54.5 (1966)
1966–67	9.8	4.5 (1967)	55.0	59.3 (1967)
1967–68	8.6	0.7 (1968)		
Mean	7.4	2.0	44.1	45.3
Correlation coefficient with time	0.27	0.47		

To avoid complications due to peak flow in the winter, the 12 monthly periods are listed for July–June instead of calendar years. March–May periods are also given as the time when the most fertiliser nitrogen will be applied to arable crops. The quantity of nitrate carried by the river is much less than the quantity of fertiliser nitrogen used. The difference is even more pronounced in the March–May period when the risks of leaching fertiliser nitrogen would be greatest. The nitrate carried by the river during the period is on average only a little more than one-quarter of the annual total. There is no evidence from this information to indicate that nitrate losses are greatest when fertilisers are most vulnerable to leaching losses.

Tomlinson concluded from his survey that while there was evidence that the concentration of nitrates tended to rise in most of the rivers he investigated, the rise was not very pronounced and there was certainly no evidence of dramatic and widespread increases in either the mean or the maximum rates of concentration. However, levels in certain rivers like the Stour have risen considerably and the position must be watched carefully and continuously. It would also be very useful to extend the survey on rivers to underground waters to see that the nitrate levels in this supply of drinking water are not increasing.

Thus, while at present the problem of fertiliser nutrients does not appear to cause immediate concern, what of the future? First, there is no evidence to suggest that potassium is causing any serious problems but little work has been carried out and a survey on some of our rivers, similar to that carried out by Tomlinson, would be of considerable value. With phosphorus, the evidence all suggests that the amount reaching watercourses from agricultural sources is small and any problems are associated more with sewage disposal than with fertilisers. It therefore would seem that any fertiliser problems are, in effect, nitrogen problems.

At present most arable crops now receive about optimum levels of nitrogen for maximum growth. The levels on grassland, however, can be expected to increase. As the amount of nitrogen lost from grassland is considerably less than that from arable land, this extra nitrogen may not pose any serious problems. However, a bigger acreage of arable crops could be grown and perhaps increase the nitrate in the drainage. New varieties of cereals may also be developed which would respond to much larger dressings of fertiliser nitrogen and heavier applications might increase the risk of pollution. Apart from normal good husbandry practices, such as minimal levels of nitrate nitrogen applications in the autumn and no long-term fallowing of land especially in the summer and early autumn, what remedies could be introduced if the position became critical?

It is doubtful if either organic or slow-acting nitrogen fertilisers would be very effective in controlling pollution, since anything which adds to soil fertility is a potential cause of nitrate formation. However, they should prevent the rapid loss of nitrate which could occur if water-soluble fertiliser were applied immediately before very heavy rain. There are, at present, some organic chemicals on the market which inhibit the nitrification of the ammonium ion. It is not known if this effect would last long enough to be useful, but nevertheless this is a possible potential method for the prevention of nitrate loss.

Other methods have been suggested by St. Amant and Beck (1970) working at the Californian Department of Water Resources USA. Among the methods they are investigating are:

1. *Algal stripping*. This entails growing a dense crop of algae in reservoirs so that at harvesting a considerable proportion of the nitrate nitrogen in the water could be removed in the algal bodies. This method is very dependent on temperature and it is unlikely that it would be very effective under the average temperature prevailing in the UK for the greater part of the year. Also, this method could produce problems in that, unless carefully controlled, algal growth could increase explosively under certain conditions and 'blooms' would result that, for example, could kill other aquatic organisms or block filters.
2. *Bacterial denitrification*. Anaerobic denitrifying bacteria which reduce nitrates to gaseous nitrogen are added to water, either free floating in a deep pond or reservoir or preferably attached to a filter at the bottom. In both cases the bacteria would need an organic source of energy and this would have to be applied in calculated amounts for effective control. This method is also temperature dependent but probably has a greater possible potential than the previous method.

Other possible suggested methods include electrodialysis, reverse osmosis and ion exchange, but these techniques are at present speculative. If the position ever becomes so acute that detailed and complex methods have to be used to purify our water supplies, they will obviously be costly and ultimately this cost will have to be borne, however indirectly, by the consumer.

CONCLUSIONS

It does not appear that either potassium or phosphorus applied in fertilisers is having any serious effects on the natural environment. The evidence presented with regard to nitrogen would indicate that the levels in water, while rising, do not at present give rise for serious concern and the point has not been reached where they are a general health hazard. However, certain river and well-water has occasionally high nitrate levels and a careful and constant watch will have to be maintained. In addition, care and attention will have to be given to the disposal of the large amounts of farmyard manure and slurries from the increasing numbers of intensive animal units.

Apart from the infrequent times when a spring application of fertiliser nitrogen may be followed by very heavy rain, the problem of high nitrate levels in drainage water is not so much one of fertilisers as of soil fertility, since all fertile soils are a potential source of nitrate loss, especially after ploughing. Nevertheless, fertilisers are an essential part of soil fertility. Looking to the future, as rates of fertiliser application are tending to level off on crops where the risk of nitrate loss is most serious, the position may not markedly deteriorate.

Thus, we do not appear to be on the brink of a major catastrophe and we do have time to set about answering long-term questions. A number of suggestions have been made both to the farmer and the water authorities, which might help to alleviate the position if the need arises.

REFERENCES

Anon. (1970). 'Losses of plant nutrients in drainage water', *Great House Experimental Husbandry Farm Guide*, 1970. 37–39

Cooke, G. W. and Williams, R. J. B. (1970). 'Losses of nitrogen and phosphorus from agricultural land', *Wat. Treat. Examination*. Vol. 19, 253–274

Gasser, J. K. R. and Bloomfield, C. (1955). 'The mobilization of phosphorus in waterlogged soils', *J. Soil Sci.* Vol. 6, 219–232

Hendrick, J. and Welsh, H. D. (1938). 'Further results from the Craibstone Drain Gauge', *Trans. R. Highld Agric. Soc. Scot.* Vol. 50, 184–202

Lawes, J. B., Gilbert, J. H. and Warington, R. (1882). 'The amount and composition of the rain and drainage waters collected at Rothamsted', *Jl. R. Agric. Soc.* Vol. 18, 1–24

Low, A. J. and Armitage, E. R. (1970). 'The composition of the leachate through cropped and uncropped soils in lysimeters compared with that of the rain', *Pl. Soil*. Vol. 33, 393–411

Miller, N. H. J. (1906). 'The amount and composition of the drainage through unmanured and uncropped land, Barnfield, Rothamsted', *J. Agric. Sci. Camb.* Vol. 1, 377–399

Owens, M. (1970). 'Nutrient balances in rivers', *Wat. Treat. Examination*. Vol. 19, 239–247

Russell, E. J. and Richards, E. H. (1920). 'The washing out of nitrates by drainage water from uncropped and unmanured land', *J. Agric. Sci. Camb.* Vol. 10, 22–43

St. Amant, P. P. and Beck, L. A. (1970). 'Methods of removing nitrates from water', *J. Agric. Fd. Chem.* Vol. 18, 785–788

Tomlinson, T. E. (1970). 'Trends in nitrate concentrations in English rivers, and fertiliser use', *Wat. Treat. Examination*. Vol. 19, 277–289

DISCUSSION

Sheen: Are there any figures available for trace elements, especially heavy metals, leaching out of soils, particularly in cases where sewage sludge has been applied as fertiliser?

Armitage: I do not know of any results for trace element losses in drainage water. Market garden crops which received heavy dressings of sewage sludge have been grown at Woburn. There was some increase in the uptake of trace elements, but the implications of these increases are uncertain.

O'Connell: *Ministry of Agriculture, Fisheries and Food.* Many studies of fertiliser losses have been carried out on drainage schemes. Using these or shallow lysimeters, the pathway from rainfall to open water has been very much shortened. There have been big nitrate nitrogen problems with the river Stour which runs through large areas of under-drainage schemes in Essex. Could under-drainage be contributing to the nitrogen pollution of rivers by shortening natural pathways? Is there a case for not under-draining a lot of low-lying areas near rivers?

Armitage: The problem of why much higher levels of nitrates are lost from lysimeters and field drains than actually appears in rivers is a very interesting one. This may be due, in part, to the uptake of nitrate by aquatic plants.

O'Connell: I think that denitrification will occur in the soil through which the water has to drain and, I believe, denitrification can account for between 15 and 40 per cent of nitrate fertiliser losses. Such losses will be reduced and the level of nitrates leached out increased by under-drainage where water passes more quickly through and out of the drains and into rivers.

Armitage: I agree that under field conditions denitrification could account for a considerable part of the nitrate losses. Also, any system of under-drainage which increased the rate of percolation through the soil would tend to reduce these losses and thereby increase the amount of nitrate in the open water. At Jealott's Hill, in the nitrogen balance of our lysimeter experiment, there were larger losses of nitrogen from the topsoil than were occurring in the drainage. We concluded that this was partly due to denitrification.

O'Connell: Free-living nitrogen fixers are only important in poorly drained soils. Losses from freely drained soils are about 15–20 per cent; on heavier basic soils and heavy organic soils, anything up to 40–50 per cent can be lost by denitrification.

Six

The use of antibiotics and artificial hormones in agriculture

T. B. Miller
Division of Agricultural Chemistry and Biochemistry,
School of Agriculture, Aberdeen

SUMMARY

Antibiotics have been found to have beneficial effects in rearing of farm animals, giving improvements in growth rates, food conversion and carcase quality. This chapter discusses their possible modes of action and the evidence for possible deleterious effects on human health via breeding of resistant micro-organisms. The Report and Recommendations of the Swann Committee set up to examine the resistance problem are critically reviewed. Resultant changes in antibiotic usage and the use of compounds without therapeutic basis ('feed antibiotics', copper and arsenicals) are described.

Synthetic oestrogen hormones have also been found to be beneficial in animal husbandry and their use and possible disadvantages to man are discussed. The present and future use of antibiotics and hormones in NE. Scotland is described to demonstrate the type of practices employed in the UK.

INTRODUCTION

Antibiotics and hormones are two types of feed additives which are administered to fattening animals for the purpose of growth promotion. The administration is either by incorporation in the diet, by parenteral injection or (in the case of hormones) by subcutaneous implantation. They are generally referred to as growth promoters which may be defined as 'any substance, other than protein, fat or carbohydrate, minerals or vitamins given in normal physiological amounts, which produces an increase in growth or feed efficiency in a higher animal over and above that normal for the species, sex, strain, age and environment' (Mudd, 1971).

USES OF ANTIBIOTICS

An antibiotic is generally defined as a compound produced wholly or partially by a micro-organism (fungus or bacterium) which can inhibit growth of, or destroy, other micro-organisms.

Antibiotics in agriculture are used with animals for the following purposes:

1. Veterinary therapy.
2. Disease prevention.
   (a) with animals showing overt signs of disease
   (b) with animals at risk but showing no symptoms
3. Growth promoting.

With 2(a) both sick and healthy animals receive treatment with therapeutic levels of an antibiotic, for example diarrhoea in pigs often associated with *Escherichia coli*, pneumonia in calves, summer mastitis in cattle. Infected and healthy animals would be treated in all cases. With 2(b) amounts of antibiotics are usually given below the recommended therapeutic dose when animals are 'stressed'.

The growth promoting properties of antibiotics were discovered by chance in 1946 following the isolation of vitamin B_{12}. Suitable sources of the vitamin were sought for the supplementation of all vegetable diets for pigs and poultry, and were found in the residues from the industrial production of streptomycin and chlortetracycline. The crude residue produced a greater growth in chicks than vitamin B_{12} alone, however, and the growth promoting effect of chlortetracycline and vitamin B_{12} was as great as the fermentation product when the chicks were maintained on diets deficient in animal protein. Similar results were obtained with other antibiotics including penicillin, bacitracin, streptomycin and oxytetracycline. Combinations of antibiotics did not produce a more favourable response than when employed singly. Moreover, there were species differences with respect to responses of individual antibiotics—penicillin, for instance, is capable of promoting the growth of pigs and poultry but not of calves, whereas tetracyclines may increase growth rates in all three species. For the purposes of this chapter consideration of the uses of antibiotics will be confined to growth promotion.

Antibiotics in poultry feeding

Most studies have shown that if an improvement in growth occurs when newly-hatched chicks are given diets containing antibiotics, the effect is greatest in the first two weeks. Increased weight gain of treated chicks ranged from 10 to 26 per cent over controls during the first 14 days and from 2 to 9 per cent during the second 14 days (White-Stevens, 1957).

This early growth stimulation is important with broilers where the feeding period is 8–10 weeks but the results are less important with birds reared to maturity for egg production. The improvement in growth is of 5–10 per cent and improvement in food utilisation of up to 5 per cent. Since the cost of food is about 80 per cent of the total cost of broiler production, the improvement in efficiency is considerable.

In recent years genetic improvements and increased nutritional knowledge have increased production and food conversion to about the same order as that attributable to antibiotics.

Antibiotics in pig feeding

The results with weanling pigs have been similar to those with broiler chicks. Early work produced a marked response at low (2–10 ppm) dietary levels of antibiotics; 10–20 per cent increase in growth rate and 5–10 per cent improvement in food utilisation have been recorded (Robinson, 1968). Since then responses have declined due to general improvements in the performance of control animals in experiments.

The greatest response is obtained during the period from weaning (15 kg) to about 45 kg weight. Up to 45 kg the animals given 4 ppm penicillin in feed produced 17 per cent improvement in liveweight gain compared with 4 per cent improvement between 45 and 95 kg (Robinson *et al.*, 1954). When feed intakes of controls and treated animals are the same, the antibiotic has been shown to produce improved food utilisation without adverse effects on carcase quality.

In a wide range of diets studied in Denmark, Clausen (1961) found that the overall response in growth rate amounts to 5 per cent with 4 per cent improvement in food utilisation. With a skim milk/cereal diet the mean responses were only 2–3 per cent but levels of 11 per cent were recorded when milk was replaced by other protein foods (meat and bone, soya bean or sunflower meal) or the cereal replaced by other carbohydrate foods (potato silage, sugar beet) or with both substitutions made.

Antibiotics in feeding cattle

The main use of antibiotics in calf rearing is in the control of potential pathogens, though they can also have a role in the young ruminant by increasing food intake. Non-maternal rearing of calves reached a high level of development before the introduction of antibiotics and where calves are home-bred, colostrum is generous and the subsequent milk substitute is of good quality, the building well ventilated and overcrowding of calves avoided, the use of such drugs is superfluous. Unfortunately, all these conditions are rarely met and where calves are not brought quickly from their place of birth to the site of rearing, routine feeding of antibiotics may be a necessity for keeping calves profitably (Roy, 1970).

In some environmental and nutritional situations, chlortetracycline and oxytetracycline have a definite beneficial effect in reducing mortality and increasing the growth rate of calves. Colostrum may not be available to provide the essential antibodies for instance, and 250 mg chlortetracyline given orally daily for the first 5 days of life followed by 125 mg/day for the next 16 days reduces diarrhoea and gives protection from death in most cases (Ingram *et al.*, 1958).

Tetracycline antibiotics are particularly useful in dealing with *Escherichia coli* infection with purchased calves. Synthetic penicillin (ampicillin) is particularly useful in the prophylaxis against salmonellosis because it is rapidly absorbed from the gut. In the case of feeding young ruminants, improved food intake, food conversion and liveweight gain can be achieved by including chlortetracycline at a concentration of 20 ppm. These improvements have been attributed to a reduction in the rate of fermentation and gas production in the rumen causing a greater proportion of the diet to be absorbed in the abomasum and intestine (Preston, McLeod and Dinda, 1959). Higher blood glucose levels in ruminant calves receiving chlortetracycline support this theory.

MODE OF ACTION OF ANTIBIOTICS

The use of antibiotics as a feed additive is based on empirical findings from feeding trials. However, the mechanism or mechanisms involved presents an intriguing problem for which the solutions are still based on speculation. There have been many proposals put forward to account for the stimulation of growth and these can be divided into those which would act directly or indirectly (Luckey, 1960).

Indirect action on intestinal microflora

Increasing 'good' organisms and decreasing 'bad' organisms

In this category one can speculate on the possibility of enhancing growth and population of organisms which synthesise nutrients, for example vitamins. This theory is less tenable when one considers the marked response with monogastric animals, which are less dependent on bacterial synthesis than ruminants and beneficial effects with the latter have been less marked. The suppression of organisms producing toxins is perhaps more plausible though experiments by Eyssen, Desomer and Con Dijck (1957) with guinea pigs have indicated definite stimulation of detrimental organisms in the caecum. Francois and Michel (1955), on the other hand, have demonstrated reduced liberation of ammonia and amines by antibiotics in a culture obtained from pigs.

Reduction of sub-clinical infection

Chicks reared in clean surroundings were found to grow better than chicks in old quarters and the latter responded well to antibiotics (Coates *et al.*, 1952). Rearing the chicks in sterile plastic cages resulted in no response to antibiotics unless they were inoculated with faecal material from chicks in 'dirty' quarters. The inoculation reduced growth rate and feeding antibiotics overcame this inhibitory effect. Increasing the level of antibiotics to 220 ppm did not result in any increased growth rate of birds in clean quarters. Similar results have been obtained with pigs though Hill and Larson (1955) found that pigs obtained by hysterectomy and raised without contact with other pigs grew at a faster rate when fed chlortetracycline than a group fed no antibiotics. Studies with calves have elicited a greater response to antibiotics in clean premises.

Direct action of antibiotics

Changes in the intestinal wall are seen when no growth stimulation or change in the flora is found, indicating a possible systemic action. Experiments where antibiotics were injected have given either negative or less positive results than from oral administration.

Several authors have reported improvement in growth with antibiotics deactivated by autoclaving, which indicates that antibiotics can act directly in the absence of bactericidal action by the compound. Penicillin deactivated by heat, penicillinase or heavy metals produced growth stimulation in pigs when administered by mouth or injection and it should be noted that the material had no antibiotic activity.

Although the precise mode of action of antibiotics in growth promotion has not been satisfactorily elucidated, it is evident that these compounds act as biological stabilisers in animals (Luckey 1960). Their effect is greatest under conditions of mild stress, for example, birth, semi-starvation, bacteriologically dirty environment, excessive production of eggs, cold weather, dietary deficiency or presence of infectious disease.

HAZARDS OF FEEDING ANTIBIOTICS

Humans who are allergic to antibiotics may have severe even fatal reactions with antibiotic treatment even with minute doses. Some antibiotics given orally can kill all bacteria normally present in the large intestine, thus causing likelihood of invasion by virulent organisms. Use of antibiotics may also lead to appearance of bacteria resistant to the drug, rendering the antibiotic useless. The consumption of animal flesh or milk from treated animals is accompanied by a potential risk to humans who may be allergic or who may become infected by resistant strains. Moreover, persons whose occupation brings them in contact with the uses of antibiotics with animals are at particular risk.

The possible hazards associated with the use of antibiotics for growth promotion have caused increasing concern in this country during recent years. In 1968 HM Government appointed the Joint Committee under the chairmanship of Professor M. M. Swann 'to obtain information about the present and prospective uses of antibiotics in animal husbandry and veterinary medicine with particular reference to the phenomenon of drug resistance, to consider the implications for animal husbandry and also for human and animal health, and to make recommendations' (Report, 1969—commonly called the 'Swann Report').

The committee reported that there was a dramatic rise in the numbers of strains of enteric bacteria of animal origin which show resistance to one or more antibiotics and that there was ample and incontrovertible evidence that such organisms were ingested by humans. They expressed particular concern that enteric organisms of the salmonella group, for example *Salmonella typhimurium*, showed a tendency to give rise to generalised infection in man. The possible development of resistance to antibiotics (for example, chloramphenicol) by such organisms could endanger human life.

RECOMMENDATIONS OF THE SWANN COMMITTEE

The committee concluded that if antibiotics are used unwisely there could be rapid propagation of resistant organisms. They refuted the statement that twenty years of experience goes to show that there are no serious ill-effects from giving antibiotics to animals.

They recommended that antibiotics without prescription for adding to animal foods should be restricted to those which:

1. Are of economic value in livestock production under UK farming conditions.
2. Have little or no application as therapeutic agents in man or animals.
3. Will not impair the efficacy of a prescribed therapeutic antibiotic through the development of resistant strains of organisms.

Many other recommendations and suggestions were made such as research into stress and improved methods of animal husbandry. The committee believed that similar economic benefits could be obtained with 'feed antibiotics', i.e. those conforming to 1, 2 and 3 above.

In 1969, prior to the publication of the report, farmers were allowed to include penicillin, chlortetracycline and oxytetracycline at concentrations up to 100 ppm feed for growing pigs and poultry without veterinary prescription. Permission for the unprescribed use of penicillin, tetracyclines, tylosin, most sulphonamides and four nitrofurans has now been withdrawn.

The 'feed antibiotics' include: zinc bacitracin—may be added to feeds for growing pigs and poultry and to calves and lambs up to 6 months old at 122 ppm; flavomycin at 25 ppm for poultry and 63 ppm for pigs; virginiamycin at 7 ppm for broilers. Growth stimulants which are not classed as 'feed antibiotics' include nitrovin (a guanidine derivative). It has no utility in the prevention of disease in man or animals, but has given consistent improvements in growth and has a wide margin of safety. Grofas (quinoxaline derivative) gives growth stimulation at 20 ppm for broilers (Lucas, 1972).

The changes recommended by the Swann Committee have brought considerable criticism in the Press from all quarters, including scientists who state that the evidence for development of resistant strains of organisms is far from conclusive. It is claimed that not one single annotated case of human harm has occurred during almost 20 years of including low levels of antibiotics in pig and poultry rations. In the US a task force of scientists was established in 1970 to review the use of antibiotics in animal feeds. The recommendations of the body which have been issued recently (Report, 1972) are very similar to those of the Swann Committee.

COPPER AND ARSENICALS

Although they are not antibiotics, the use of copper salts and arsenical compounds as feed additives requires consideration in the present context. Pulverised copper sulphate (at 250 ppm Cu) in feed improves both growth rate and feed conversion efficiency of pigs from weaning to slaughter by 5–8 per cent (Lucas, 1972). Growth response is least in the heavier pigs and may average as high as 22 per cent in pigs weaned at 10–28 days (Wallace, 1970).

Copper at 250 ppm of the diet represents 30 or 40 times the nutritive requirement of the animal for this element. The use of such massive amounts followed observations by Dr. Braude of Shinfield that fattening pigs have a craving for copper. They showed preference for copper when confronted with a choice of several metals and they have the ability to discriminate in favour of rations containing copper (Braude, 1945). The responses obtained with higher copper rations resembled those obtained with antibiotics. There is also considerable evidence of synergism between copper and antibiotics as feed additives (Lucas and Calder, 1957).

Copper toxicity with pigs has occurred at 500 ppm and occasionally at 250 ppm (Lucas, 1964). Protection against toxicity is afforded by the inclusion of adequate levels of iron and zinc (Suttle and Mills, 1966). With cattle, and particularly with sheep, toxicity of copper, is a more serious problem. Sheep can absorb copper very efficiently and toxic levels are reached when the diet contains 10–20 ppm. The contamination of sheep diets with copper is therefore a serious problem. Moreover, the disposal of pig manure from animals on high copper diets can contaminate soil and pasture causing death of sheep. The aerobic degradation of pig waste is also impaired by the presence of copper (Robinson, Draper and Gelman, 1971). High dietary levels of copper in the pig results in very high storage in pig liver thus rendering liver and other offal unfit for human consumption.

In 1905 Ehrlich discovered the therapeutic action of arsenicals. There followed considerable interest and a number of compounds including sodium arsanilate, though the discovery of sulphonamides and other chemotherapeutic agents followed later by antibiotics diminished the interest in arsenicals.

The growth promoting properties of 3-nitro-4-hydroxyphenyl arsonic acid were noted in 1946 following studies of the therapeutic properties against caecal coccidiosis in chicks (Morehouse and

Mayfield, 1946). Later, stimulation of growth in pigs was demonstrated (Carpenter, 1951).

Following the Swann report there was renewed interest in arsenicals to replace antibiotics in feeds. Arsanilic acid and 3-nitro-4-hydroxyphenyl arsonic acid added at 30–100 ppm to diets for growing chicks and pigs, and sometimes up to 300 ppm for short periods, are most frequently used to promote growth. Responses are similar to those obtainable with antibiotics or copper. Arsenical compounds have been associated with improved utilisation of nutrients by reducing heat loss (Sibbald and Slinger, 1963). With 100 ppm arsanilic acid, Barber, Braude and Mitchell (1971) showed improved growth rate and feed conversion efficiency in pigs of 6 per cent and 2 per cent respectively. Performance was not improved further with the inclusion of 250 ppm copper though the levels of copper in the liver were reduced (Barber *et al.*, 1971). The levels of arsenic in the liver are markedly reduced by withdrawing arsenic from the ration for six days. Regulations in this country require withdrawal of arsenicals for some days before slaughter, but the rule is difficult to enforce.

USE OF ARTIFICIAL HORMONES

In the 1940s and 1950s numerous experiments were conducted on both sides of the Atlantic to demonstrate the action of a number of hormone substitutes. The materials were either injected, fed or implanted. Of all the materials examined only diethylstilboestrol (DES) and hexoestrol have retained a role in ruminant production systems though male androgenic steroids are used with pigs (Beacom, 1963).

The commercially significant effects with ruminants were an increase in live weight and lean tissue with a concomitant reduction in fat. Administration to fattening lambs improved rate of gain, better carcase appearance and better feed efficiency (see *Table 6.1*).

Table 6.1 INCREASES IN LIVE WEIGHT AND CERTAIN CONSTITUENTS OF THE CARCASE OF LAMBS DURING FATTENING (*Source: Preston, Gee and Crichton, 1957*)

	Hexoestrol group (lb)	*Control group* (lb)	*Change over control* (%)
Increase in live weight	52.0	41.0	+27
Increase in fat	10.7	11.4	− 6
Increase in protein	2.9	2.1	+38
Increase in water	10.6	7.0	+51
Increase in bone	2.3	1.0	+130

The application of results of empirical findings from early experiments was rapid despite a general ignorance of the mode of action of oestrogens. The development of radioimmunoassay techniques for the estimation of protein hormones from the pituitary has done much to elucidate the mode of action and the extent of metabolic control that can be exerted.

With ruminants the effect of hormone treatment on the castrated male animal has been well documented. The response with different sexes is in the order: castrated male>male>female (Swan, 1970). It has been observed that stilboestrol increases voluntary food consumption and animals with access to high-quality foods make the greatest liveweight gain.

Oestrogens cause a surge in the secretion of growth hormone coupled with increased plasma insulin (Hafs, Purchas and Pearson, 1971). Growth hormone influences the number of functional ribosomes in muscle tissues, causing an increase in the uptake of amino acids and their incorporation in intercellular protein. The overall effect is manifested by greater nitrogen retention and reduced blood urea levels (Hutcheson and Preston, 1970). Recent studies with male rats by Lloyd *et al.* (1971) have shown increased mitotic activity in the pituitary, increased pituitary weight and increased serum growth hormone levels, rising from 20 ng/ml to 342 ng/ml after nine days.

Although meat production can be improved by drug administration, it must be recognised that some of the drugs may be retained by the tissues, thus affecting humans consuming the meat. Naturally occurring oestrogens are as potent as the stilbene oestrogens when administered parenterally. When orally administered, however, the latter are much more active than steroidal oestrogens because of the reduced hepatic inactivation of the synthetic compounds (Pincus, 1948). This accounts for the detection of stilbene residues in tissues of treated animals. Hexoestrol is used as a growth stimulant in the UK in preference to stilboestrol because it produces less pronounced side effects than the latter (Hammond, 1957).

There is considerable disagreement in the reporting of oestrogenic residues which may be attributed to differences in dose level, interval between treatment and slaughter and the type of tissue assayed. Early reports of residual oestrogenic activity in beef were based on assays of tissues following administration of high dose levels, for example 120 mg stilboestrol. Oestrogen residues are not detected in the tissues of steers treated with 24–36 mg stilboestrol which is the presently recommended range of levels in the USA (O'Mary, Cullison and Carman, 1959). Oral doses of 10 mg per dose are not detected in tissues of steers and a 48 h withdrawal period is recommended before slaughter (Gossett, Smith and

Downing, 1956). Kidney and liver show the highest levels. Implantation in the ear has been the favoured method of administration in this country.

Some concern has been expressed on the possible accumulation of hormones in the soil and subsequent appearance in plants from the use of excreta from hormone-treated animals as manure. Oestrogens eliminated in the faeces and urine of hormone treated animals may occur in either free or bound form. Studies on the stability of hexoestrol in soil and its uptake by plants suggest that, for this oestrogen, it is unlikely that the oestrogenic content of plants would be increased as a result of fertilising the soil with excreta from treated stock (Glascock and Jones, 1961).

Side effects of oestrogenic treatment include development of mammary glands, enlargement of secondary sex glands and prolapse of rectum or vagina which has caused serious death losses in certain groups of lambs (Jordan, 1953), although implants of 12 mg have not produced such deleterious effects (Lamming, 1956). Similar adverse reports have been made in relation to pigs by Dinusson, Klosterman and Buchassan (1951).

PRESENT AND FUTURE USE OF ANTIBIOTICS AND HORMONES IN NE. SCOTLAND

In NE. Scotland, pig producers who used antibiotics before the Swann Report have now changed over to acceptable feed additives such as grofas. Most breeding units use 200 ppm copper for fattening rations only, but in specialised fattening units, where pigs are bought-in, the inclusion of growth-promoting additives, such as grofas, is preferred.

Arsenicals are used in some fattening units. These tend to be mixed on the farm and there is some evidence of improper mixing which has caused arsenic poisoning. Here the difficulty of enforcing regulations concerning the withdrawal before slaughter causes some concern. Zinc bacitracin is used in turkey feeds generally, though this compound may also be used for calves.

Hormone usage is less widespread. The new product 'Maxymin' is becoming increasingly popular in fattening pigs. This is a mixture of methyl testosterone and DES which is sold as a premix. The product is included in fattening rations between 20 and 40 kg liveweight until 72 h before slaughter. The withdrawal period is necessary to ensure that no residues of hormones are present in the carcase or offal. Liveweight gains have been increased by 3.6 per cent with increases in food conversion rates of 2.3 per cent. The subcutaneous fat was reduced by 10 per cent thus improving the grading of carcases (Elanco, 1970). Maxymin is also used along with grofas for fattening pigs.

The use of hexoestrol was widespread with cattle about 10–15 years ago. Hexoestrol was generally implanted in the ear in doses of 60–120 mg. The introduction of barley beef systems encouraged the use further, because of the beneficial effects of reducing fat in the carcase. The hormone is particularly beneficial with early maturing cattle, such as Aberdeen Angus. Recently, however, late-maturing dairy breeds, for example Friesian and Ayrshire, have been crossed with a Hereford bull, which produces late-maturing offspring. The desired effect by the farmer is to achieve 'finish' with beef × dairy crosses by adding a layer of subcutaneous fat, and hexoestrol militates against this development.

ACKNOWLEDGEMENTS

The author wishes to acknowledge the assistance of Dr. V. Fowler and Miss K. Elliott of the Rowett Research Institute for the provision of information relating to hormones.

REFERENCES

Barber, R. S., Braude, R. and Mitchell, K. G. (1971). 'Arsanilic acid, sodium salicylate and bromide salts as potential growth stimulants for pigs receiving diets with and without copper sulphate', *Br. J. Nutr.* Vol. 25, 381

Beacom, S. E. (1963). 'The effect of diethylstilbestrol and estradiol-testosterone implants on rate and efficiency of gain and on carcass quality of market pigs fed different finishing diets', *Can. J. Anim. Sci.* Vol. 43, 374–384

Braude, R. (1945). 'Some observations on the need for copper in the diet of fattening pigs', *J. agric. Sci. Camb.* Vol. 35, 163

Carpenter, I. E. (1951). 'The effect of 3-nitro-4-hydroxyphenyl arsonic acid in the growth of swine', *Arch. Biochem. & Biophys.* Vol. 32, 181

Clausen, H. (1961). 'Bericht ubes die Vortragstagung', *Neue Erkenntnisse in der Verfutterung Von Sofaschrot*, November 1961

Coates, M. E., Dickinson, C. D., Harrison, G. F., Kon, S. K., Porter, J. W. G., Cummings, S. H. and Cuthbertson, W. F. J. (1952). 'A mode of action of antibiotics in chick nutrition', *J. Sci. Fd. Agric.* Vol. 3, 43–48

Dinusson, W. E., Klosterman, E. W. and Buchassan, M. L. (1951). 'Stilbestrol, effect of subcutaneous implantation on growing-fattening swine', *J. Anim. Sci.* Vol. 10, 885–888

Dodds, E. C., Goldberg, L., Lawson, W. and Robinson, R. (1938). 'Oestrogenic activity of certain synthetic compounds', *Nature, Lond.* Vol. 141, 247

Elanco (1970). *Maxymin literature*, Elanco Products Ltd. Broadway House, Wimbledon, London

Eyssen, H., Desomer, P. and Con Dijck, P. (1957). 'Further studies on antibiotic toxicity in guinea pigs', *Antibiotics Chemother.* Vol. 7, 55–64

Francois, C. and Michel, M. (1955). 'Action de la penicilline et de l'aureomycine sur les proprietes desammantes de la flore intestinale du porc', *C.R. Acad. Sci.* Vol. 240, 124–126

Glascock, R. F. and Jones, H. E. H. (1961). 'The uptake of hexoestrol by plants and its persistance in soil', *J. Endocrinol.* Vol. 21, 373

Gossett, F. O., Smith, F. A. and Downing, J. F. (1956). In *Proceedings of the Symposium on Medicated Feeds*, Washington D.C., 1956. New York; Medical Encyclopoedia Inc.

Hafs, H. D., Purchas, R. W. and Pearson, A. M. (1971). 'A review: relationships of some hormones to growth and carcass quality of ruminants', *J. Anim. Sci.* Vol. 33, 64–71

Hammond, J. (1957). 'Hormones in meat production', *Outl. Agric.* Vol. 1, 230

Hill, E. G. and Larson, N. L. (1955). 'Effect of chlortetracycline supplementation on growth and feed utilization of unsuckled baby pigs obtained by hysterectomy', *J. Anim. Sci.* Vol. 14, 1116–1121

Hutcheson, D. P. and Preston, R. L. (1970). 'Stability of diethylstilbestrol and its effect on performance in lambs', *J. Anim. Sci.* Vol. 32, 146–151

Ingram, P. L., Shillan, K. W. G., Hawkins, G. M. and Roy, J. H. B. (1958). 'The nutritive value of colostrum for the calf. 14. Further studies on the effect of antibiotics on the performance of colostrum-deprived calves. *Br. J. Nutr.* Vol. 12, 203

Jordan, R. M. (1953). 'Effect of stilboestrol on suckling and fattening lambs', *J. Anim. Sci.* Vol. 12, 670–679

Lamming, G. E. (1956). 'The use of hormones in meat production', *Agric. Prog.* Vol. 32, 31–38.

Lloyd, H. M., Meares, J. D., Jacobi, J. and Thomas, F. J. (1971). 'Effects of stilboestrol on growth hormone secretion and pituitary cell proliferation in the male rat', *J. Endocrinol.* Vol. 51, 473–481

Lucas, I. A. M. (1964). 'Modern methods of pig nutrition: the impact of recent research', *Vet. Rec.* Vol. 76, 101

Lucas, I. A. M. (1972). 'The use of antibiotics as feed additives for farm animals', *Proc. Nutr. Soc.* Vol. 31, 1–8

Lucas, I. A. M. and Calder, A. F. C. (1957). 'Antibiotics and a high level of copper sulphate in rations for growing bacon pigs', *J. agric. Sci., Camb.* Vol. 49, 184

Luckey, T. D. (1960). *Antibiotics—their Chemical and non-Medical Uses*, ed. H. S. Goldberg. New York; D. Van Nostrand Co. Inc.

Morehouse, N. F. and Mayfield, O. J. (1946). 'The effect of some aryl arsonic acids on experimental coccidiosis infection in chickens', *J. Parasit.* Vol. 32, 20–24

Mudd, A. J. (1971). British Council Course for Pig Specialists held at R. College Vet. Surgeons, London

O'Mary, C. C., Cullison, A. E. and Carman, J. L. (1959). 'Implanted and oral stilboestrol for fattening steers', *J. Anim. Sci.* Vol. 18, 14

Pincus, G. (1948). *The Hormones*, New York; Academic Press

Preston, T. R., Gee, I. and Crichton, J. A. (1957). 'Hexoestrol and carcase quality', *Agric. Rev.* Vol. 111, 39–45

Preston, T. R., McLeod, N. A. and Dinda, P. K. (1959). 'The effect of chlortetracycline on growth of early-weaned calves', *Anim. Prod.* Vol. 1, 13

Report, 1969. *Joint Committee on the use of Antibiotics in Animal Husbandry and Veterinary Medicine*, Cmnd. 4190, London; HMSO

Report, 1972. US Department of Health, Education and Welfare. Federal Register, February 1972, 37, FR 2444

Robinson, K. L. (1968). *Antibiotics in Agriculture*, Proceedings of the 5th Symposium of the Group of European Nutritionists in Jouy en Josas, 1966

Robinson, K. L., Coey, W. E. and Burnett, G. S. (1954). 'The use of antibiotics in the food of fattening pigs', *J. Sci. Fd Agric.* Vol. 5, 541

Robinson, K., Draper, S. R. and Gelman, A. L. (1971). 'Biodegradation of pig waste: breakdown of soluble nitrogen compounds and the effect of copper', *Environ. Pollut.* Vol. 2, 49

Roy, J. H. B. (1970). *The Calf*, Vol. 2, 3rd edn. London; Iliffe Books

Sibbald, I. R. and Slinger, S. J. (1963). 'The effects of breed, sex and arsenicals and nutrient density on the utilization of dietary energy', *Poult. Sci.* Vol. 42, 1325–1332

Suttle, N. F. and Mills, C. F. (1966). 'Studies of the toxicity of copper to pigs. 1. Effects of oral supplements of zinc and iron salts on the development of copper toxicosis. 2. Effect of protein source and other dietary components on the response to high and moderate intakes of copper', *Br. J. Nutr.* Vol. 20, 135

Swan, H. (1970). Proc. 25th Am. Texas Nutr. Conf., 128–130

Wallace, H. D. (1970). 'Biological responses to antibacterial feed additives in diets of meat producing animals', *J. Anim. Sci.* Vol. 31, 1118

White-Stevens, R. H. (1957). 'Antibiotics as dietary supplements for poultry', *Vet. Rec.* Vol. 69, 217

DISCUSSION

Knights: How much will changes in antibiotic usage following the Swann Report cost farmers?

Miller: In the early press reports on the Swann Committee Recommendations, it was stated that this would cost farmers about £35 million per annum: how they arrived at these figures, I am not sure.

Mason: This figure was not calculated by farmers—their estimate was about £2–3 million per annum. It is very difficult to assess the costs to the industry and the country as a whole.

Fitzgerald: *May & Baker Ltd.* What will be our position following entry into the Common Market?

Mason: Our regulations on the uses of antibiotics are tighter than those obtaining at present in EEC countries. Feeding of hormones and arsenicals are, however, covered under present EEC regulations and so in this respect, we may be improving our ways if we enter the Community and accept these regulations.

With respect to changes in antibiotic usage, the recommendations made have been based on the possibility of transfer of resistant organisms to humans without the Swann Committee ever having proved or demonstrated such transfer. On the other hand, the Swann Report conveniently ignores the fact that some 85 per cent of the antibiotics used in this country are used directly on humans and the possibility of the production of resistant organisms is very much greater here.

The speaker did, and very rightly so in my opinion, draw attention to the value of antibiotics at time of stress. Now that they can only be administered at therapeutic levels as prescribed by a vet, there is a danger that farmers will use less antibiotics at this time to avoid consultation and prescription delays. More disease will result and more antibiotics will have to be used later for therapeutic purposes—such factors make the assessing of costs very difficult.

O'Connell: What sort of levels of arsenicals are normally fed to animals? Has any work been done regarding the passage of arsenicals through the digestive system and their emergence onto grass which may then be ingested by cows and the arsenicals perhaps passed on to humans in milk or meat?

Miller: It is often difficult to decide whether figures for levels of arsenicals fed to animals quoted in the literature refer to whole organic complexes or to arsenic itself. We have fed chickens at 34 ppm elemental arsenic and this did provide appreciable quantities in the droppings. In relation to pigs, I am not sure how much passes out in the faeces or what residues are likely to be found on pasture. I feel the risks associated with misuse of arsenicals are far greater than those with antibiotics.

Mason: It is possible that copper may build up in pig manure used as fertiliser and could this be dangerous to sheep since they are very sensitive to copper? Is there not in fact a danger from copper sulphate pasture dressings used to control snail vectors of liverfluke parasites? If one used pig manure containing copper, could not one complete two jobs at once?

Miller: Copper is normally most toxic in the free ionic form and, in pig droppings, most of it will be chelated and some bound in insoluble form. Therefore, though there may be 100 ppm in the material, a large proportion is innocuous. Pig manure would therefore probably not be effective against snails.

Seven

The effects on man of the use of chemicals in agriculture

M. Sharratt*
Department of Health and Social Security

SUMMARY

Many chemicals used in agriculture could, if care is not taken, adversely affect human health. The difference between the toxicity of such chemicals (their innate ability to cause injury) and the hazard they present to health (the probability that they will cause injury) must be appreciated. These factors are discussed in relation to the possible toxic hazard of agrochemicals on humans, the assessment of hazards to human health and the interpretation of the results of testing procedures and the actions taken to reduce hazards to a minimum.

It is concluded that the voluntary testing and control schemes currently operating in the UK have proved adequate in ensuring that the hazard to human health is extremely low in relation to the benefits to be gained from the use of agrochemicals.

INTRODUCTION

Agrochemicals can have both beneficial and harmful effects on health. Some of the benefits to be gained from their use have been described in previous chapters, for example fertilisers help directly to increase crop yields, as do pesticides, herbicides and fungicides indirectly. The health of farm animals and therefore the supply and quality of their meat are also improved by the use of pesticides (for example in sheep dips). Hormones and antibiotics are included in very small quantities in the feed of some animals to improve their

* The views expressed here are those of the author. They do not necessarily reflect the views of the Department of Health and Social Security.

growth rates and feed conversion rates. Many chemicals are used for other purposes which farmers and horticulturalists find helpful, from the simple disinfectant to chemicals which have more specific activity against mites, slugs, snails and eelworms or which remove leaves or dry foliage to facilitate harvesting, for example the haulm killers used on potatoes. Chemicals can be used to increase or decrease the rates of crop growth and, particularly in horticulture, they are valuable as soil sterilants to control pests, diseases and weeds. Cereals, grassland, vegetables, fruit, hops, glasshouse crops and mushrooms may all receive treatment with agricultural chemicals and farm animals and poultry are treated to remove infections and infestations. Most people would now agree that the use of agricultural and other chemicals on the farm has increased the availability and decreased the cost of food and so helped to provide the needs of an ever-increasing world population. Since good food is essential to health and cheap food allows larger numbers to get a fair share, it is obvious that agrochemicals play a significant role in promoting world health. However, fears have frequently been expressed that the increasing use of agrochemicals constitutes a hazard to our health, and this chapter discusses, not these benefits, but the potential health risks and the work that is done to ensure the safety-in-use of chemicals in agriculture.

Many chemicals used in agriculture are undoubtedly highly toxic and may, if care is not taken, be hazardous to health. However, even if a chemical is highly toxic it may not be hazardous, while chemicals of low toxicity may, under certain circumstances, present highly significant health hazards. These apparent anomalies should be explained.

TOXICITY

The toxicity of a chemical is its innate ability to cause injury to living things. The toxicity of a chemical is therefore the sum of the various untoward effects it can have under a variety of circumstances on the human or animal body. The toxicity of, for example, the pesticide malathion is as follows: 50 per cent of a group of rats will be killed if each one is given between 1 and 2 g of the substance for every kg of body weight. If it is given by injection into the peritoneal cavity, less than half this amount will kill 50 per cent of the group. It kills by inhibiting the enzyme cholinesterase which occurs in the nervous system; this inhibition leads to malfunctioning of the nerve transmission system in the body and eventually to death as described in Chapter 2. Malathion is hydrolysed in the body of animals and man to less toxic substances. It reduces the growth rate when fed at high dosage levels to rats, mice and chicks and also decreases the amount of cholinesterase which is found in their red blood cells. Rats fed for 2 years on a diet containing malathion, so that each animal received up to 250 mg for each kg of its body weight every day, develop no more cancers than do control animals. When 8 mg was given daily for 32 days or 16 mg daily for 47 days to human volunteers, nothing untoward happened. When the dose was increased to 24 mg a day for 56 days, the plasma cholinesterase decreased although there was no clinically detectable ill-effect (FAO/WHO, 1968).

This catalogue of untoward effects is somewhat different from what many understand by toxicity. Further, since it is possible to cause some sort of injury with all chemicals, if exposure is at a high enough dosage level or by a particular route, it must be true that all substances are toxic. It is perhaps difficult to think of such substances as water and oxygen as being anything other than absolutely harmless chemicals. However, even these common substances, which do no apparent harm and are, in fact, essential to life, can cause injury under certain circumstances. Water, for instance, can be highly dangerous to people with certain kidney diseases and even to normal healthy people if taken in large quantities when they are salt-deficient; such a state may occur in the tropics. A number of years ago there was an outbreak of a disease (retrolental fibroplasia) in which a fibrous layer formed behind the lens of the eye. The disease occurred in children who, because of premature birth, had been put into incubators into which pure oxygen was passed to assist respiration. The disease was traced to the use of pure oxygen and when this was replaced by oxygen-enriched air, no more cases occurred. Since all chemicals can cause injury and must therefore be considered toxic, it is obviously wrong for chemicals to be referred to as 'toxic' or 'non-toxic'; all chemicals should be rated on a continuous scale from very low to very high toxicity. Much confusion has arisen because of the failure to distinguish between toxicity and hazard.

HAZARD

The hazard presented to living things by a chemical of given toxicity depends not only on its toxicity but also on the circumstances under which it is used and whether or not these uses lead to a significant exposure to the chemical.

Hazard is best defined as the probability that a substance of given toxicity will cause injury under a given set of circumstances. It is obvious that a highly toxic substance is much more likely to be hazardous than a substance of very low toxicity if both are used in the same way. However, substances with the same toxicity might, because of their physical properties, present entirely different hazards. For example, potassium cyanide and hydrogen cyanide have very similar toxicological properties but potassium cyanide is much less of a hazard than hydrogen cyanide. Because hydrogen cyanide is absorbed by the skin, is volatile and rapidly absorbed through the lungs, it rapidly leads to deaths if released into the atmosphere while the non-volatile potassium salt is used extensively in industry with no harm to health. While the toxicity of a compound is important, it is only one factor to be considered when assessing the hazard presented by the use of a chemical in agriculture; the numerous other factors which control the total exposure and frequency of exposure of individuals to the chemical are of equal importance.

Many highly toxic chemicals are used for agricultural purposes and although, by careful control (Bates, see Chapter 8), the exposure of individuals is such that the hazard from their use is extremely low, accidents do happen. Organomercury compounds are valuable for combating seed-borne diseases and they are put on to cereals in the form of seed dressings. When the mercury compound and the treated seeds are properly handled they present no risks to health. However, it was reported recently (Damluji and Tikriti, 1972) that there was an outbreak of poisoning from the mercurial compound ethyl mercury *p*-toluene sulphonanilide in Iraq. The poisoning occurred among farmers whose wheat grain had been dressed with the fungicide. The number of cases admitted to hospital was more than 5500 and some 280 people died. A similar outbreak on a smaller scale occurred in that country in 1961 (Jalili and Abbasi, 1961; Damluji, 1962). Unsown mercury-treated seed had apparently been sold illicitly for human consumption and farmers had also eaten the seed left after sowing. Another case of poisoning from an organomercury compound occurred in a family in New Mexico. The family killed one of their pigs and stored the meat in a deep freeze. The animal was apparently healthy although at the time it was killed some of the others of the herd were showing signs of muscular incoordination, an indication of methyl mercury poisoning. Two months after the pig was killed the three children in the family became seriously ill and eventually methyl mercury poisoning was diagnosed. Eight months later they were still undergoing treatment. For several weeks before being killed the pig had been fed mainly on millet seed swept from the floor and given away by the local granary; the seed had been treated with mercury compounds and there were considerable residues which had accumulated in the pork (Curley *et al.*, 1971). Methyl mercury compounds are of high toxicity. When they are properly used they are of very low hazard but these cases show that when they are misused they present a high degree of hazard.

There have also been several cases of poisoning from the valuable herbicide paraquat; these again illustrate the importance of distinguishing between toxicity and hazard. Paraquat is a highly charged molecule and therefore poorly absorbed through the body surface or the gut and it is rapidly excreted from the body in the urine. It is a local irritant and if splashed into the eye or onto the skin will cause irritation and inflammation and sometimes blistering. It can cause finger-nails to be shed and also delays healing. When particles are inhaled they can cause nose bleeds and when swallowed, a solution causes sore throat, vomiting, abdominal pain and diarrhoea. Another toxic effect of great interest follows ingestion of paraquat. Although the patient may survive the acute irritative effects, severe lung damage develops over a period of 10 days to 3 weeks and this is rapidly progressive and fatal. Between 1953 and the present time no deaths have occurred in persons occupationally exposed to paraquat. A few cases of trivial poisoning are reported each year (MAFF, 1961–65) but these consist of only minor rashes or the effects of a splash in the eye or on the skin. From this point of view, paraquat is no more of a problem than the many other irritant chemicals used in the household or factories. Although the hazard of the chemical as used by agricultural workers therefore is extremely low, people have been killed by it. Deaths have occurred in adults (Hargreave, Gresham and Karayannopoulos, 1971) and children (McDonagh and Martin, 1970) who have consumed various quantities. In general, death has been due to the lung damage. All deaths have, in fact, been from suicide or from drinking paraquat from unlabelled or mis-labelled bottles in which it had been stored. The hazard from paraquat in unlabelled bottles is very great indeed but, as was stated in Parliament, '. . . as with all cases that have come to our notice nothing untoward would have happened if the prominent warnings on the manufacturer's containers had been observed' (*Hansard*, 1971–72).

POSSIBLE TOXIC ACTION OF AGRICULTURAL CHEMICALS

In order to assess the hazard of using a chemical, its toxicological properties must be investigated thoroughly and the design of the tests depends to a large extent on the manner in which man is likely to be exposed. During the preparation of chemicals for application and while the actual application is proceeding, the chemical dust or vapour might be inhaled by workers. It might also get onto their skin or into their eyes and/or be accidentally ingested. Workers may also be contaminated with the chemical when carrying out other crop treatments or harvesting crops; here again skin contact is the likely mode of contact. The general public, however, is exposed in a different way, that is to very small amounts of the chemical which might occur as residues in the crops or the meat of treated animals. While exposure of workers is generally to high concentrations of the chemical for short periods, the general public tends to be exposed to very low concentrations for long periods; for example, a chemical which is used with great success to treat cereals may be ingested in the diet in small amounts every day by large segments of the population for several years or even the whole of a life span. Thus tests have to be carried out to determine both the acute effects of small and large doses of the chemical when administered by mouth, by inhalation and to the body surface, and also the long-term effects of ingesting very small amounts regularly.

In designing tests it is also necessary to take account of all the possible types of injury which could occur. Chemicals can cause many different types of injury. It is not possible to discuss them all, but many fall into groups. Some chemicals damage the various surfaces of the body—the skin, lungs, eyes and alimentary tract—by direct contact; paraquat and diquat are good examples but there are many others. Other substances cause no injury on direct contact but the internal organs are affected following their absorption into the body. Some substances have to be concentrated in particular organs before they cause injury; inorganic mercury, for instance, concentrates in the kidney and methyl mercury in the brain. Other chemicals injure by their pharmacological effects so that, for example, the organophosphorus and carbamate insecticides act by depressing cholinesterase activity and affect nerve transmission. The effects of some chemicals can be detected by biochemical techniques. For example, several are known to increase the metabolising capacity of the liver by increasing the activity of microsomal enzymes. This may be advantageous in some cases since rapid metabolism can reduce the carcinogenicity of certain substances (Alfred and Gelboin, 1967), but equally it can increase the toxicity of some chemicals. Carbon tetrachloride is much more toxic to rats with a high microsomal enzyme activity than to normal rats. The main problem with a few substances is that, over a long period of time, they accumulate in tissues. Examples include cadmium, lead and, on the agricultural side, several organochlorine insecticides including DDT. Toxicologists also have to remember that some chemicals may injure by causing cancer or by producing malformations of the newborn. Others may also act as mutagens and damage the genetic material of the somatic, or the reproductive, cells. Other types of injury which are of particular importance in relation to agrochemicals can be grouped under the heading of irritation and allergic sensitisation, and tests have to be designed to eliminate chemicals which might cause serious outbreaks of asthma, urticaria, eczema and other forms of dermatitis.

TOXICITY TESTS

To examine the possibility that such forms of injury might occur entails extensive laboratory tests—the following account is merely an outline of some of these. The rationale of testing and the techniques used are described in more detail in publications by WHO (1956; 1962; 1967), the Council of Europe (1967) and MAFF (1971). Before starting tests it must be certain that the substance to be tested is the one to which workers will be exposed and the one which occurs as a residue in food. Much attention is therefore paid to specification of the commercial product and to the identification of residues of the chemical occurring in foodstuffs. Investigations are carried out to examine the absorption of the chemical from the gastrointestinal tract, its distribution in the body tissues, its excretion and the possibility that it might be concentrated in a specific organ. The metabolites produced in plants and animals may also be examined in a similar way. It is often necessary in these studies to use radioactively labelled compounds. Where the metabolites have to be traced in human volunteers their use can present difficulties but it is probable that carbon ^{13}C will be a great help in the future. Tests on the acute toxicity are carried out on several animal species in single and multiple doses to give an idea of how much chemical it might take to injure or to kill exposed humans. In these tests the chemicals are administered by several routes including by mouth, on the skin and, if necessary, by inhalation. A figure called the LD_{50} is calculated; this dose, which kills 50 per cent of the group of animals, is valuable for comparative purposes. The cause of death is also investigated; it is essential to understand this in order to know whether the lethal dose in animals is really relevant to man. For example, a substance that kills by inhibiting an important enzyme is much more likely to be a hazard to man than a substance which kills animals by causing gastro-intestinal irritation when high concentrations are administered. A multiple dose test gives an indication of what is likely to happen to man following multiple exposures over a short period, such as might occur in agricultural practice. Particular attention is paid to the detection of substances which might adversely affect the skin and eyes.

Short-term studies, in which animals receive daily doses of a chemical for about one-tenth of their life-span, are carried out to determine the possible effects of longer-term exposure. These tests are the equivalent of several years exposure in humans. Rats, mice and other species of animals are fed or injected with the chemical daily for several months and they are looked after and examined exactly as are hospital patients. They are observed carefully several times a day, weighed periodically, their food intake is measured and biochemical and haematological tests done to assess their state of health. The only difference from patients is that at the end of the test they are killed by anaesthetic and post-mortem examinations carried out, their organs weighed and histological examinations made of all of their organs and tissues. Sometimes the tests are modified so that animals which are adversely affected are allowed to recover or are given treatments for poisoning in order to determine the way to treat humans accidentally over-exposed to the chemical. Long-term tests are then performed in many cases. These are to study the effects of substances that might accumulate or produce their effects slowly over an even longer period. The most important toxic effect investigated is the carcinogenic potential. Rats and mice are the most convenient species to use and they receive treatment for most of their life-span—this amounts to about two years in the case of the rat. During the test the body weight and food intake are determined, the blood and urine are analysed and organ function tests performed. The illnesses each animal suffers and its survival time are recorded and when killed at the end of the tests, the organ weights, the number of tumours and the incidence of other pathological lesions are compared with those of untreated animals. In addition to demonstrating carcinogenic potential the test indicates what other toxic effects are likely to occur during these periods, these periods being equivalent to those over which the general population might be exposed to small amounts of the chemical occurring as residues in food. Other tests are, if necessary, carried out to determine the effects of the chemical on reproductive ability or to assess the teratogenic or mutagenic properties. If it is an organophosphorus compound, chronic nerve damage is investigated by a special test (MAFF, 1971) because some organophosphorus compounds have been known to cause delayed nervous damage. When a particular chemical is used frequently with another chemical, tests may be carried out for synergistic activity of the two chemicals together.

INTERPRETATION OF THE RESULTS OF TESTS

The results of all of these tests on animals indicate only what will happen to the particular animals exposed. One fact of toxicology which raises all sorts of problems for the toxicologist is that all species of animals and all animals in any species do not necessarily react quantitatively or qualitatively in the same way to the same chemical substances. This means that the only species of animal which is absolutely perfect for testing the likely effects on man of agricultural chemicals (or inded any other chemicals) is man himself. Tests cannot be carried out on human volunteers in the early stages and animals have to be used as models. Fortunately, many animal species react in a similar way to man to a wide variety of chemicals, and rats, mice, dogs and monkeys and many other species are used as models when testing agricultural chemicals. The results of these animal tests are interpreted in terms of the likely effect of the chemical on man by considering known similarities and differences in anatomy, physiology, biochemistry, pharmacology and pathology between experimental animals and man. The results of investigations into the mechanisms of action of the chemicals are also taken into account as are those results of special tests carried out on volunteers after completion of the animal tests or results of accidental exposure during the manufacture or use of the material. Such tests may include comparison of the metabolites produced by the test species and by man.

Having decided, in qualitative terms, the ill-effects to be expected in human subjects who might be exposed to a chemical, it is necessary to consider quantitative aspects. This is particularly important where long-term ingestion of small residues in food is involved. In general, humans tend to be more sensitive than animals to many chemicals. In addition, any two people do not necessarily react in the same way to a given amount of chemical; a dose which is harmless to one person may make another very sick indeed. When it comes to transferring quantitative animal data to humans, it is necessary to take these factors into account and, unless a large volume of experience is available on its actions in man, to allow very wide safety margins between the safe animal dose and calculated 'safe' human dose. The use of a large safety factor also reduces the health risks for another reason; for all substances there is some definite level at or below which a substance can be taken into the body without health being harmed in any way. A highly toxic chemical will, if taken in large quantities, lead to injury and perhaps death but if the dose is reduced sufficiently it can be taken regularly without health being harmed in any way. Every day highly toxic chemicals are consumed in the diet; potatoes contain the alkaloid solanine which, when eaten in large amounts, can cause vomiting, diarrhoea and abdominal pain. The small amounts taken daily in potatoes have no detectable ill-effect on the large number of people who consume them. Rhubarb and spinach contain oxalic acid; a large dose of oxalic acid will lead to kidney damage and death whereas the smaller amounts we take in our food are perfectly harmless. Tea and coffee contain caffeine; again no harm ensues from the quantity we take in these beverages. However, as little as 50 times that amount is estimated to be the human lethal dose. This concept is of obvious importance when agricultural chemicals are considered because, although many are highly toxic and could cause injuries if taken in large quantities, no injuries to health will follow if it is ensured that the intake is sufficiently low.

In practice, it has been found that if the dose of chemical shown to be harmless to animals when given over a long period (for example two years in the rat) is divided by a safety factor of about 100, this intake level is safe to humans when taken daily in their diet. This extensive and expensive testing procedure therefore helps to decide what amount of chemical residue occurring in food will not harm humans. A great deal more information is needed in order to assess the hazard to the consumer arising from the use of an agrochemical. Much of this will be described in Chapter 8. It is necessary to determine, for example, the amounts of chemical residue occurring in various treated foods under various conditions of use and this depends to some extent on how long before harvesting or the slaughter of the animal treatment is carried out. The scale of use is also important; there will be a difference in the degree of health hazard between a few pounds of a chemical to be used in a desert area and many tons to be used near crowded cities. When it is decided what crops or animals will be treated with an agrochemical it is possible, knowing the residues, to calculate the likely intake of the chemical by the average human being. If this is below the calculated safe human long-term intake level, the use of the chemical is likely to be accepted. It is extremely unlikely at present that any chemical found to be a carcinogen or a mutagen could be accepted at any level in the diet.

Although the residues in food of a particular chemical may be found to be acceptably low, it may present such a hazard to the person applying it (because of its acute toxicity) that its use in agriculture could not be considered. The tests, however, indicate the way in which a chemical could be dangerous to the applicator. This allows safety precautions to be worked out for those who are to apply the chemical and also the way of treating workers who are accidentally over-exposed to the chemical. With adequate safeguards, most chemicals can be made safe for operatives. These safeguards may vary from restriction of the use of the chemical to trained personnel who wear protective clothing, to labelling the container carefully with instructions on how to use the chemical, precautions to take and what to do if exposure occurs. Details of the scheme by which the safety of pesticides to workers and consumers is ensured in the UK are discussed in Chapter 8.

CONCLUSION

This country is fortunate in having a first-class agriculture industry backed by a first-class agrochemical industry, one which is highly conscious of its responsibilities not to impair the health and well-being of agricultural workers and the population at large. The larger manufacturers are often backed by their own toxicological laboratories, developed and run at great expense to study the safety aspects of products and the problems of residues in foodstuffs. The close co-operation of the industry with the Ministry of Agriculture, Fisheries and Food and the Department of Health and Social Security in the safety precautions scheme to be described later in Chapter 8 is of great advantage to this country. The very responsible attitude shown by the industry and this co-operation ensures that both the farmer and consumer can benefit greatly from the careful and controlled use of agrochemicals without there being more than minimal risks to health. Abuse or misuse of agrochemicals can still occur despite the recommendations arrived at by government and industry and it is unlikely that legislation can prevent this. Education of the farmer and the general public on hazards involved in the use of chemicals on farms and home gardens and on the need for strict observance of manufacturers' instructions could possibly help a great deal in this respect.

REFERENCES

Council of Europe (1967) *Agricultural Pesticides*, 2nd edn. Strasbourg

Alfred, L. J. and Gelboin, H. V. (1967), 'Benzpyrene hydroxylase induction by polycyclic hydrocarbons in hamster embryonic cells grown in vitro', *Science*. Vol. 157, 75

Curley, A., Sedlak, V. A., Girling, E. F., Hawk, R. E., Barthel, W. F., Pierce, P. E. and Likosky, W. H. (1971). 'Organic mercury identified as the cause of poisoning in humans and hogs', *Science*. Vol. 172, 425

Damluji, S. F. (1962). 'Mercurial poisoning with the fungicide Granosan–M', *J. Facul. Med., Bagdad*. Vol. 4, 83

Damluji, S. F. and Tikriti, S. (1972). 'Mercury poisoning from wheat', *Br. Med. J.* Vol. 1, 804

FAO/WHO (1968). *Evaluations of some pesticide residues in food*, FAO/P1. 1967/7/11/1 WHO/Food Add./68.30

Hansard (1971–72). Vol. 827, 261

Hargreave, T. B., Gresham, G. A. and Karayannopoulos, S. (1969). 'Paraquat poisoning', *Postgrad. Med. J.* Vol. 45, 633

Jalili, M. A. and Abbasi, A. H. (1961). 'Poisoning by ethyl mercury toluenesulphanilide', *Br. J. Ind. Med.* Vol. 18, 303

McDonagh, B. J. and Martin, J. (1970). 'Paraquat poisoning in children', *Arch. Dis. Child.* Vol. 45, 425

Ministry of Agriculture, Fisheries and Food (1961–65). *Reports on Safety, Health, Welfare and Wages in Agriculture*, 1961–1965. London; HMSO
Ministry of Agriculture, Fisheries and Food (1971). *Pesticide Safety Precautions Scheme agreed between Government Departments and Industry*. London; HMSO
WHO (1956). *Toxic Hazards of Pesticides to Man*. Technical Report Series No. 114. Geneva; WHO
WHO (1962). *Principles governing consumer safety in relation to Pesticide Residues*. Technical Report Series No. 240. Geneva; WHO
WHO (1967). *Procedures for investigating inte::tional and unintentional food additives*. Technical Report Series No. 348. Geneva; WHO

DISCUSSION

Reay: When you are judging human safety levels from experiments on rats working on the LD_{50} figure, do you take into consideration the actual slope of the probit line being used to derive the LD_{50}? The safety factor of 100 you mention may be fine if the slope is steep but quite the opposite if it is not.

Sharratt: I personally take little notice of the probit line because the LD_{50} is difficult to interpret in terms of human toxicity. It indicates merely how large an overdose has to be given to kill the animal. If it is known whether the LD_{50} lies in the range 0–5, 5–50 or 50–500 mg/kg or above, one can derive a very approximate idea of the likely human lethal dose. What matters far more is the mode of toxic action. This is often very difficult to determine but more time should be spent on studying this aspect in many cases.

Knights: How much research into toxicity and hazards is done by government departments like yours and how much by the producer himself? Don't problems arise when a manufacturer wants to keep the formulation of a new product secret? This certainly causes problems in waste disposal by contractors from industrial processes.

Sharratt: The manufacturers do most of the work themselves and do it extremely well. Smaller manufacturers have to send it out to commercial toxicology laboratories. No toxicological work is done by the government but the assessment of results is carried out by scientific committees on which suitably experienced people sit. (This will be discussed in Chapter 8.) BIBRA does a lot of toxicological work but the emphasis there is more on basic problems and food additives. There are no problems with commercial secrets—all data remain confidential to the assessing committees and even other government departments are not allowed to see them.

Puri: It is worrying to know that everything we eat is toxic to some extent and one would like to know just how safe is 'safe'! You say wise and careful use of potentially toxic substances will not lead to harm yet other people hold much more alarmist views. Is this due to insufficient dialogue between agrochemical producers and environmentalists?

Sharratt: If a doctor examines a patient, he eventually assesses his condition in terms of the quality of life he leads. If you look at the nation with clinicians' eyes, the position is not so bad—life expectancy is increasing, we have better food, more time for leisure and so on. There are potential problems with agrochemical usage but the purpose of the safety scheme to be outlined later and the large amount of research going on in this area is to tackle them before they arise. Many people do, in my opinion, take an alarmist attitude. Of course it provides a good story for the mass media who tend to give publicity to well-meaning but often ill-informed people who worry a lot about the environment. Many get worried because they don't really understand the problems and do not realise that they are being tackled. I think it is a good thing that these people do make themselves heard since it keeps us on our toes.

Puri: Three-quarters of the world population are not even aware of these problems, yet worries in the developed countries have given rise to restrictions on some chemicals and to things such as 'Blueprint for Survival'.

Sharratt: An example of this is the case of DDT. In the UK, we don't need to use large amounts. DDT and its metabolites have been found to accumulate in the body fat, although this doesn't seem to do us any harm. However, because of this it was agreed voluntarily to reduce its use to the minimum possible. But it is wrong that other countries should ban DDT just because Britain and the USA do so, and for a country with a malaria problem to ban it at the present time would surely be a disaster.

Eight

Legislation regarding the use of chemicals in agriculture

J. A. R. Bates
Plant Pathology Laboratory, Ministry of Agriculture, Fisheries and Food

SUMMARY

Because of the possible adverse effects they may have on non-target organisms, the introduction and use of agrochemicals must be controlled in some way to reduce risks associated with their use to a minimum. Most developed countries exercise such control via some form of registration scheme and the functioning and effectiveness of the Pesticides Safety Precautions Scheme and the Agricultural Chemicals Approval Scheme as currently operated in the UK are outlined. Mention is also made of the extent to which existing legislation covers various aspects of agrochemical usage and a brief summary of the provisions of relevant acts is given in an appendix.

The effects of the review of safety arrangements in the UK made by the Government's Advisory Committee and the Recommendations of the Royal Commission on Environmental Pollution are discussed in relation to possible future changes in legislation. Legislation in other countries and possible future changes, especially with regard to harmonisation of legislation of EEC countries and the UK, are also discussed.

INTRODUCTION

Modern pesticides are important in the production of food but since many of them are toxic to non-target organisms as well as target organisms, as described in earlier chapters, they carry problems of health hazards to the persons who handle them and apply them as well as to the consumers of treated crops and other foodstuffs containing residues. They may also, of course, have adverse effects on wildlife. Because of their potential toxicity to non-target organisms, they are frequently in the news and much has been written and said about them, ranging from the wildly inaccurate and highly sensational to the factually correct and dull. However, since pesticides may have far reaching effects on man and wildlife, this study would be incomplete if it did not include some account of the measures taken to ensure that pesticides put on the market are safe enough to use.

How best to reduce the hazards of pesticides to men and animals is a problem that has occupied many individuals and organisations the world over. Most developed countries control the introduction of pesticides through some type of registration scheme and although the basic principles employed in various countries may not be very different, there are considerable differences in the ways in which laws are used to control the use of pesticides. The discussion in this chapter will centre mainly around the Pesticides Safety Precautions Scheme as operated in the UK and existing legislation of relevance to pesticide usage, together with a discussion of possible future changes in national and international legislation, especially with regards to the EEC.

Whilst this chapter concentrates on controls of the use of pesticides within the wide definition of the terms given under the Safety Precautions Scheme (see below), some details of controls of the use of veterinary antibiotics currently in operation in the UK are given in Chapter 6.

THE PESTICIDES SAFETY PRECAUTIONS SCHEME

In the UK, at present, control is achieved through the voluntary Pesticides Safety Precautions Scheme which has been operating since 1954, although it formally began in 1957. This scheme, originally called the Notification of Pesticides Scheme, was devised as a result of close co-operation between government departments and the pesticides industry and this co-operation has been, and still is, a major factor in its continuing success.

The government's main source of advice on risks arising from pesticides is the Advisory Committee on Pesticides and other Toxic Chemicals. This committee has an independent chairman and other independent members from outside government circles, together with representatives, both administrative and technical, of interested departments and research councils. There are no trade representatives on this committee.

The Scheme is designed to safeguard human beings (whether they be users, consumers of treated crops or other members of the public) livestock, domestic animals and wildlife against risks from pesticides. For this purpose, manufacturers, distributors and importers who propose to introduce new pesticides or new uses for pesticides undertake to notify the government of their intention, before they are introduced. The Scheme applies to all chemicals formulated as pesticides; that is insecticides, fungicides, herbicides, growth regulators, rodenticides and similar products used in agriculture, forestry, horticulture, home gardens and food storage in the UK. It covers:

1. Products based on a new active ingredient, that is a chemical not previously used as a pesticide in the UK.
2. Any extension of the use of an existing pesticide, for example, from non-edible to edible crops, or to additional crops; to home garden use; from outdoor to indoor use or vice-versa or even a new distributor of an existing product.
3. Changes in the way of using an existing pesticide that could produce a new or increased risk, for example, changes in formulation or rates and methods of application.

Notification is not normally expected while the product is at the stage of laboratory or small-scale trials carried out by the firm's staff but if the product is to be used by agricultural or horticultural workers, or if treated crops are to be available for human or animal consumption, then the product must first be notified. Each notification is, in fact, a request from a firm for official agreement to its proposals and the firm must justify its request by submitting suitable supporting information.

The firm provides all the information needed to enable the government to advise on the precautions which should be taken when the pesticide is used and it agrees not to introduce new products until agreement has been reached on appropriate precautions. These, together with the name of the active ingredient, are then included on the label of every container of the product offered for sale. Although the Pesticides Safety Precautions Scheme is still voluntary, regulations under the Farm and Garden Chemicals Act, 1967 require, by law, the name of the active ingredient to be on the label (see the Appendix to this chapter for a summary of this and other relevant Acts governing various aspects of the supply and use of pesticides).

Notifications concerned with agriculture, forestry, horticulture and home gardens are made to the Ministry of Agriculture, Fisheries and Food at the Plant Pathology Laboratory, Harpenden and those concerned with food storage, domestic and animal husbandry uses to the Pest Infestation Control Laboratory, Slough.

Guidance to firms on the amount and type of data required is given by the Scheme through its many appendices and working documents. For example, advice is given on the scope of toxicological studies required, presentation of data on pesticide residues in crops and standard tests for the toxicity of pesticides to fish and bees and effects on wildlife. The information required includes full details of the composition of the product, its proposed method of use, mode of action, toxicity, persistence and other data relevant to its safe use.

Procedure

The technical secretariat of the Pesticides Safety Precautions Scheme can process notifications in one of several ways and those involving trials with new chemicals or minor changes in use and labels of chemicals already on the market can be cleared quickly by using a panel of advisers. In other cases notifications go through the committee procedure. Such notifications go first to the Scientific Subcommittee of the Advisory Committee which carefully scrutinises the supporting technical information and assesses its value.

The Subcommittee, composed of scientists with expert pesticide knowledge, includes chemists, biologists, entomologists, toxicologists, a plant pathologist and a veterinary expert as well as members with a special interest in wildlife. This membership allows the Committee to take into account all aspects likely to arise from the use of a pesticide and to agree recommendations. Not infrequently a notification is rejected, usually on the grounds of insufficient information, in which case it may be resubmitted at a later date.

Whichever course is taken, final clearance depends on decisions by government departments. Any recommendations by advisers or the Subcommittee are based purely on a scientific assessment of the notification and in its advice to its parent Advisory Committee the Scientific Subcommittee is concerned only with the scientific aspects of the safe use of pesticides. The Advisory Committee then takes into account factors other than those of a purely scientific nature.

The government also decides whether or not a chemical is toxic enough to be included in the Agriculture (Poisonous Substances) Act, 1952 and its Regulations. The Act is to protect employees from poisoning by the more dangerous compounds, by ensuring that they are supplied with and use suitable protective equipment. Provisions of the Act are restricted to employees of farmers, growers and contractors, and obligations are imposed on the employer, who must provide the prescribed clothing and make sure his workers wear it.

The Regulations take into account the fact that one method of using a chemical may be more dangerous than another. Thus, other things being equal, soil application is safer than normal spraying which, in turn, is safer than the use of aerosols. The Ministry issues a leaflet APS/1 entitled 'The Safe Use of Pesticides on the Farm' which gives a summary in non-legal terms of the main provisions of the Regulations as well as much other general advice.

Of the 250 or so agrochemicals generally available in this country, only about 30 are considered sufficiently toxic to need regulating. Chemicals included in the Act are usually also subject to the provisions of the Pharmacy and Poisons Act, 1933 and regulations made under it which restrict sales and impose certain labelling requirements and conditions under which listed poisons may be bought, packed, and stored on shop premises.

If a chemical is not sufficiently toxic to require regulating, advice on such user precautions as are necessary will be given, using the wording the manufacturer will be required to put on his label. Consultation with industry has produced a labelling guide which is part of the Scheme and this ensures that the advice going to the user in official recommendations and from industry through the label is the same.

In this country it is believed that the most effective and practical way of protecting the consumer is to lay down a code of practice so that the user, in following advice, will know that his harvested crop will not contain harmful pesticide residues. Official recommendations on the safe use of a pesticide on growing crops contain limiting conditions such as:

1. Naming the crop on which the pesticide may be used.
2. The maximum rate and frequency of application.
3. The minimum interval which must elapse between last application of the pesticide and harvesting the crop. This is the most important of all.

The official advice is so designed that, provided the chemical is used properly, there will be either no residue or any residue present will be at a level acceptable to expert medical opinion.

In spite of unfavourable publicity the reader must not imagine that every foodstuff contains residues of the pesticide applied. The proportion of our foodstuff treated with pesticides is quite modest and even if a crop has been treated, it does not follow that there will be a residue at harvest. In many instances the normal commercial use of a pesticide is such that no trace is left at harvest and in only a small proportion of cases is there a residue at all in a harvested crop.

Official studies of the problem of pesticide residues in foodstuffs in the UK are conducted under the direction of the Panel on Residues of Pesticides in Foodstuffs (1966–67; 1966–69; 1970–71). Selective surveys of home produced and imported foods have been carried out since 1961 and the results published. Studies on total diet have also been made and, in general, as expected, residues if detectable are very low and the results justify to a large extent the Panel's practice of selecting for analysis specific foods for special consideration. The work is continuing.

Protecting wildlife

The third group at risk from the use of pesticides is a very mixed one and includes livestock, game, birds, fish, bees and other wildlife. The Scheme usually requires information on the toxicity of a chemical to birds, bees and fish as well as the data on various experimental animals. Not all pesticides are potentially harmful to wildlife but even so, each use of a chemical is carefully assessed, and specific advice on the protection of wildlife is given when necessary.

In recent years there has been growing appreciation of the significance of the contamination of the environment with persistent pesticides and information on the persistence of new chemicals is asked for. Draft recommendations on safe use, after consideration by the advisory committee and government, are then sent to the notifier as the official reply to his notification.

When government and firms concerned agree on recommendations they are published in a loose-leaf dossier entitled 'Chemical Compounds Used in Agriculture and Food Storage—Recommendations for Safe Use in the United Kingdom', freely available to anyone in this country or abroad. It is supplied to government departments, local authorities, medical officers of health, public libraries, hospitals and universities, industry, the farming and medical press and many other organisations and persons interested in the safe use of pesticides. Although applicable solely to the UK, several thousand copies go to over 60 overseas countries. Official interest in a chemical does not cease with the publication of recommendations for safe use. Chemicals are constantly under review and recommendations can be, and often are, revised in the light of new information. Research in all aspects of pesticides is in progress in many government departments, the Medical and Agricultural Research Councils and in universities and agricultural institutes.

As knowledge increases, official requirements become more

sophisticated and the examination of the data submitted by industry more critical. Industry recognises this continued stiffening of the official attitude towards pesticides—an attitude which also applies to other classes of chemicals, for example drugs and medicines—and accepts it as inevitable in our progress towards safer pesticides.

Safety record in the UK

Since 1950, there have been very few accidents in agriculture and horticulture due to pesticides. An analysis of accidents in agriculture since 1962, a period over which pesticide use has steadily increased, is given in *Table 8.1*. Several non-fatal accidents each year were without definite proof that pesticides were to blame. The evidence would seem to indicate that hazards from applying pesticides are well-controlled by existing methods (especially when compared with the principle single cause of agricultural deaths, namely overturning tractors, which killed an average of 30 people each year in earlier years.) Further, there have been no known cases of illness in any country resulting from pesticides residues in food when pesticides have been used according to directions.

Table 8.1 TABLE SUMMARISING NUMBERS OF FATAL AND NON-FATAL ACCIDENTS IN AGRICULTURE 1962–71 (*Source: HMSO Annual Reports—Report on Safety, Health, Welfare and Wages in Agriculture*)

Period	*Fatal accidents in agriculture*		*Non-fatal accidents in agriculture*	
	Total	*Due to pesticides*	*Total*	*Due to pesticides (and other chemicals)*
1.10.62–30.9.63	116	(1)*	12 777	26
1.10.63–30.9.64	99	—	11 866	19
1.10.64–30.9.65	87	—	10 408	27
1.10.65–30.9.66	115	—	9 514	12
1.1.66–31.12.66	113	—	9 352	9
1967	114	—	8 572	26
1968	114	—	7 387	22
1969	115	—	7 387	15
1970	105	—	6 291	30
1971	118	—	5 711	23

* The fatality in 1963 was a non-agricultural accident in which a child died

Review of safety arrangements in the UK

In 1967 the government published the Advisory Committee's 'Review of the Present Safety Arrangements for the Use of Toxic Chemicals in Agriculture and Food Storage'. Although the voluntary scheme has worked well for years and is an outstanding example of co-operation between government and industry, there is always the possibility that a hazard may arise because the present voluntary arrangements cannot, because of their nature, be comprehensive. A few small companies who are not members of trade associations and a few individuals who import products for their own use and for limited distribution to farming friends were thought to be ignoring the scheme.

This 'loophole' was one of the reasons given for the Advisory Committee's recommendation that a mandatory scheme should replace the existing voluntary one. This change, which was supported by the Association of British Manufacturers of Agricultural Chemicals (now British Agrochemicals Association) whose members already honour the voluntary scheme, if carried, would have meant that it would be an offence to sell, supply or import a pesticide product which has not been licensed by the government.

Proposed legislation in the UK

The proposed legislation, which was circulated for comment to interested parties, would, if adopted, control the supply and labelling of pesticide products. The licensing authority would have powers to control the storage and misuse of certain pesticides and to impose residue limits for some active ingredients in certain crops. It was proposed that labelling should be controlled as part of the licence, and regulations would control advertising and prevent the use of objectionable and misleading names for pesticide products.

A subject in which thoughts and conditions are constantly changing, it is the flexibility of the present scheme which is perhaps prized most and the proposals said that it was essential to preserve the flexibility of the present scheme in respect of information required before a licence can be granted. The new Pesticide Bill proposed to include the current provisions of the Agriculture (Poisonous Substances) Act, 1952 and the Farm and Garden Chemicals Act, 1967. The present requirements under the Agriculture (Poisonous Substances) Regulations for certain records to be kept of the use of scheduled pesticide products by employees would be extended to cover employers and self-employed persons and the records themselves would be expanded considerably to include date of purchase of the product and quantity purchased, and the details of the use or uses to which it has been put. There were also proposals to make it an offence to misuse certain pesticides in certain specific ways, for example, to scatter dressed seed as bait for wild birds or to feed dressed seed to animals or poultry.

In February 1971 the Royal Commission on Environmental Pollution supported the recommendation for legislation, but after a reassessment by the Advisory Committee and further discussion, the Minister of Agriculture announced in the spring of 1972 that the present government saw no need to proceed with legislation at the moment. This is the current position.

THE AGRICULTURAL CHEMICALS APPROVAL SCHEME

A second scheme covering the use of pesticides in the UK is the Agricultural Chemicals Approval Scheme, which in 1960 replaced the old Crop Protection Products Approval Scheme which started in 1942.

The Agricultural Chemicals Approval Scheme is operated by an organisation of the same name and is concerned solely with the efficacy of proprietary products. Products, however, cannot be considered for approval until they have first been cleared through the Pesticides Safety Precautions Scheme. While all new chemicals or new uses of existing chemicals are notified, not all the products based on these chemicals are necessarily put up for approval. The Approval Scheme is in all senses a voluntary one and charges are made for products submitted for approval.

At the present time, well over 600 proprietary products are officially approved and over 60 firms support the Approval Scheme. Nearly every chemical, of any consequence, used on crops is represented by at least one approved proprietary product. The Approval Scheme originally covered only those chemicals used for crop protection purposes in the field, glasshouse and gardens; although an extension to materials used to protect stored products is being discussed.

In seeking approval for a product, the applicant must provide full details of the formulation, the method of analysis used, and data on stability in storage, suspensibility, compatibility and other relevant information on the chemical and its physical properties. He must also provide details of the trials carried out in support of the efficacy of the product for pest, disease or weed control, and this information must tie in with the ultimate recommendations for use on the label.

The purpose of the Approval Scheme is to help growers and their advisers to select products of known performance and appropriate for the particular job in hand. Thus it would fail in its purpose if it dealt only with the backstage work and ignored the final presentation. The best of materials will be found wanting if used incorrectly and the presentation to the users of the product, that is the label, must be as clear and concise as possible. A great deal of time and effort is put into labelling both by the manufacturers and the approval organisation and agreement of the claims and directions for use are the final stage in the granting of approval. Once approval has been granted, new labels will carry the 'A' mark surmounted by a crown as a means of identification.

An approved label may not be altered without reference to the approval organisation and any additions to it or changes in the recommendations must be justified by further supporting data. Most firms revise their labels at regular intervals and the opportunity is taken on these occasions to ensure that the recommendations covering safe use as well as directions for efficient use are as up to date as possible.

As mentioned before, the schemes concerned with both pesticide safety and efficiency are voluntary and to be effective they must be suitably publicised. In the case of the Pesticides Safety Precautions Scheme this publicity is directed towards manufacturers and distributors usually through their trade organisations. With the Approval Scheme the publicity must be directed at a very large number of growers and this is achieved in one way by the publication of the Annual List of Approved Products which is available free on request. It contains not only a list of approved proprietary products but also reproduces certain information on their safe handling and use from the recommendations sheets referred to earlier.

The theme in the UK since the introduction of chemicals for pest control has been the close collaboration between the official side and industry. The pesticide industry has voluntarily accepted a good measure of control; it has agreed to the restricted use of some chemicals and even the withdrawal of several. Without this co-operation from industry the schemes could not operate and legislation would undoubtedly have already been introduced.

LEGISLATION IN OTHER COUNTRIES

Legislation on the control of pesticides is inevitably complex. This is the consequence of the many ways in which pesticides can affect human health which, in turn, results in the control of pesticides in most countries being the responsibility, not of a single authority, but of a number of authorities. Those principally concerned are usually authorities responsible for agriculture, public health and labour, but others may be involved, for example, with the transportation of pesticides and aerial spraying. In addition, various boards or committees may be set up to advise the authorities or to perform certain duties in the field of pesticide control, thus further complicating the situation. As a result, pesticide control legislation usually consists of a number of different items, each concerned with some particular aspect of the problem such as, the registration of pesticides, the licensing of manufacturers, supply and application, occupational health, aerial application or residues in food. There may also be legislation on such specialist subjects as fumigation or the dressing of seeds, or restrictions on the use of particular pesticides. All that can be attempted here is to summarise briefly what appear to be some important features of pesticide control legislation and to consider ways in which necessary harmonisation might be achieved.

Some countries exercise control over both safety in use and efficacy, while others control one or the other. In some countries, the protection of the operator stops with label directions, but in others, the law imposes responsibility on employers in respect of their employees. Many countries make use of the idea of an experimental permit, temporary clearance or licensing to allow new pesticides to be field tested and some registration authorities undertake a critical laboratory and field examination of new products. Few governments actually carry out much toxicological work. Most advanced countries take into account hazards to wildlife and at least one country has specific legislation which covers risks to bees. Again many authorities check labels to ensure that adequate guidance is given on safe and efficacious use and symbols to indicate the degree of toxicity may also be required. Common names or chemical names are generally compulsory on labels so that active ingredients can be readily identified, and medical advice on antidotes is required in some cases.

Need for harmonised legislation

Each country designs its legislation to fit into its own legal system, its economy and to some extent, to suit the national temperament of its peoples and for those countries which already have comprehensive legislation on the use of pesticides, it might present great difficulties to ask them to change to a newly devised harmonised legislation. On the other hand, in developing countries there is, at present, an opportunity to reduce the number of new laws, with the possibility of securing uniformity in such new laws and in the thinking of those who administer them. This would be desirable, but developing countries, however, usually have limited technical resources and need guidance on drafting pesticide laws.

An FAO Working Party on the Official Control of Pesticides set up in July 1963, with terms of reference which included advising on 'a model licensing and approval scheme' has proposed a graded scheme which countries could enter at a point suitable for their financial, scientific and administrative resources. Very large countries may use pesticides in vastly different climatic zones with possible different attendant hazards and a central organisation to deal with basic data such as toxicology and analytical methods is essential. Such an organisation could also co-ordinate the work of appropriate field stations operating under the different climatic conditions.

Where regulatory schemes already exist in neighbouring countries then much benefit and saving could be achieved by some form of collaboration. Western Europe offers, perhaps, the greatest opportunity for showing that such collaboration could be the basis for a Regional Pesticides Registration Scheme. Already the Benelux countries, Belgium, Netherlands and Luxembourg, have proposed certain joint legislation concerning pesticides. Other members of the EEC, France, Italy and West Germany, have their own registration schemes, but it is only a matter of time before there will be harmonisation of EEC pesticide legislation. Further evidence of the widespread appreciation of this need to harmonise legislation is the existence in Australia of a Pesticides Co-ordinator to co-ordinate the various pesticide legislations existing in the States of Australia. Certain African and Asian countries currently considering legislation could well follow these examples and introduce a regional scheme based preferably on the model law produced as a result of the FAO Working Party's Study (FAO, 1970).

Scientific requirements for registration

Most countries require that the firm developing a product must be responsible for providing the data that the national authority needs for the successful registration of the product. Before pesticides are accepted in any country considerable data on their toxicity to various species of experimental animals are required. All pesticides are toxic to some form of life, otherwise they would have no practical value and it is therefore essential that their potential toxicity to man should be studied at an early stage. Logically, a registration scheme requires the submission of adequate toxicological data to enable all hazards to be assessed and the exact scope of any study of toxicity will depend very much on the proposed use and subsequent degree of exposure of user and consumer to the chemical. Nevertheless there are certain points which must be covered in each instance. For each pesticide there is a need for certain 'basic toxicological data' which are required for the primary assessment in any country. The scope of supplementary toxicological studies which would depend on the proposed uses in a country could be agreed beforehand between the firm and the national authority concerned.

A useful first step in harmonising requirements is the booklet *Agricultural Pesticides* published in 1969 by the Council of Europe and currently being revised. This booklet gives guidance to manufacturers on the type of data to be supplied to a national authority in support of an application to market a pesticide, and covers toxicity, residues and wildlife requirements, and includes a labelling guide. There is an urgent need for some standardisation, at least of the basic data, required by different countries and it is possible that the Council of Europe guide will be adopted more widely in Western Europe in the near future.

Residues legislation

Countries differ considerably in their attitude to and control of pesticide residues in foods. Some countries, for example the USA, Canada, West Germany and Holland, have comprehensive laws defining maximum residue levels in most foodstuffs. The UK controls pesticide residues indirectly by the general provisions of the Food and Drugs Act, 1955 and such regulations as those setting limits of lead and arsenic in food, and also through the voluntary Pesticides Safety Precautions Scheme. Many developing countries restrict pesticide residues by controlling import, sale and use of pesticides. The majority of countries, however, do not yet control residues by legislation.

Some countries that have enacted comprehensive pesticide residue legislation are listed in *Table 8.2*, with a few typical tolerances to illustrate the existing divergences of opinion among pharmacologists, toxicologists and the legislative bodies around the world. Many other countries are actively considering the establishment of legislation on tolerances to control pesticide residues in their own agricultural production, both for domestic consumption and to ensure the acceptability of their product in international trade. Other countries with some tolerance restrictions are Australia, Austria, Belgium, Denmark, Finland, France, Japan, Poland and Sweden; countries actively involved in establishing pesticide residue research and evaluation centres include Argentina, Brazil, India, Norway, Spain, Thailand, the Philippines and the United Arab Republic. There is no doubt that countries which have not yet initiated any action of this sort may be forced to do so by national and international pressures from individuals, agencies, and foodstuff distributors concerned with the maintenance of public health and also through the fear that pesticide residue tolerances could serve as very effective trade barriers. It will be deplorable if tolerances, designed for consumer protection, are ultimately used in lieu of customs tariffs to prohibit importation of foods.

Table 8.2 EXAMPLES OF COUNTRIES WITH COMPREHENSIVE RESIDUES LEGISLATION SHOWING SELECTED ILLUSTRATIVE TOLERANCES (ppm) OF INSECTICIDE RESIDUES IN FOOD

Insecticide	*Canada*	*Italy*	*Japan*	*Netherlands*	*Germany*	*EEC*	*USA*	*USSR*
Carbaryl	2.0	3.0	—	3.0	3.0	3.0	10.0	—
Chlordane	0.3	0.2	—	0.1	zero	0.1*	0.3	zero
DDT	7.0	1.0	0.5 & 1.0	1.0	1.0	1.0	1.0	0.5
Malathion	4.0 & 8.0	3.0	—	3.0	0.5	—	8.0	8.0
Parathion	1.0	0.5	0.3	0.5	0.5	0.5	1.0	1.0†
Dieldrin	0.1	0.2	—	0.1	zero	0.1*	0.05	zero
Lindane	10.0	2.0	0.5	2.0	2.0	2.0	10.0	—
Aldrin	0.1	0.2	—	0.1	zero	0.1*	0.05	zero

* Individually or combined
† Including metabolites

The establishment of legally permitted amounts of pesticide residues in foodstuffs in any country depends on adequate governmental laboratories and residue analysts. Alternatively, certificates of compliance with residue regulations may be required of the producer or importer of the foodstuff, necessitating residue anlayses of consignments somewhere between production and distribution to retail markets. An alternative approach is to recommend the 'minimum interval' required between last application of the pesticide and harvesting the crop, the philosophy being that a few pilot analyses of the crops, together with experience and residue data accrued elsewhere, can be broadly applied to a particular crop/pesticide combination in the local situation to adequately protect the consumer. In general, this 'minimum interval' concept is tenable and dependable, for it is based upon the time required after application for a pesticide deposit to disappear or otherwise lose its original identity sufficiently to be well below the level accepted by toxicologists for that pesticide chemical on and in that crop. Several countries use both tolerance and 'minimum interval' requirements, whereas some other countries such as the UK currently use only the 'minimum interval' requirement.

In March 1967, the EEC Council issued a draft directive concerning the harmonisation of legislation on pesticide residues on and in foodstuffs and animal foods. This laid down tolerances for residues of certain pesticides on or in fruit and vegetables (except for potatoes). The other possible alternative approach for EEC was to attempt to harmonise the already established intervals between last application and harvesting. Since, however, the persistence of residues depends to a large extent on climatic factors and regional conditions of pest control, it was felt that the intervals were necessarily different and that harmonisation would be difficult. Moreover, safety or waiting periods are difficult to control and cannot be taken into consideration for the enforcement of regulations. The establishment of tolerances, in the EEC's view, allows the appropriate control of foodstuffs in international trade and confirmation that permissible levels of residues are not exceeded. In addition, such a system leaves individual countries free to select appropriate means for producing commodities in accordance with the EEC requirements on residues.

In arriving at the figures proposed, the EEC adopted a recommendation made in 1961 by a joint FAO/WHO working party: 'In determining how much residue should be allowed in food, the main principle to be observed is that the amount should not be higher than that which results from "good agricultural practice", provided that the final amount of residue in the daily food is not greater than the amount accepted as safe for long-term consumption by man.'

The EEC gave priority to fruits and vegetables since these commodities, at least in EEC countries, were considered to give rise to the main source of pesticide residues in the diet. In addition, more residue data were available for these classes of food than for others. Although numerous data are available, all countries may not have equal resources for analysis, apart from which member countries are not always producing the same food commodities and do not always use the same pesticides in pest control. Nevertheless, although it is not too difficult to amass data based on 'good agricultural practice' it has proved very difficult so far to reach agreement on levels. A second draft directive on residues in cereals is currently being circulated for comment.

There is a body of opinion that believes that without tolerances there can be no effective means of controlling the use of pesticides. However, as far as is known both Denmark and the UK have so far achieved adequate consumer safety in home-grown foodstuffs by methods based on advice on the proper use of pesticides including a recommended minimum interval between last application and harvest. One cannot disregard the fact that not one reported incident of illness has resulted from residues following the recommended use of pesticides under systems of control varying from the highly complicated system of legal tolerances of the USA to our own voluntary system which has no tolerances. Nevertheless, there is a trend in Europe today to change to control based on the establishment of tolerances because these are more direct and effective tools for control and enforcement.

In addition, there are moves towards the establishment of international tolerances under the aegis of the Codex Committee on Pesticide Residues (CCPR). The report of the 3rd session of the Codex Alimentarius Commission, 1965, contains a 10-step procedure for the elaboration of world-wide Codex standards. All the steps, however, are not always necessary but are subsequent to the CCPR preparing 'proposed draft provisional standard'. The CCPR has prepared a list of pesticides together with priorities for consideration by the joint FAO/WHO meeting on pesticide residues (Pesticides in the modern world, 1972—see Further Reading). Priorities have been given to those pesticides which leave substantial residues in food of importance in international trade.

In conclusion, it would appear that we, in the UK, as in most other countries, recognise that we have entered a period characterised both by a fuller understanding of the risks and advantages of pesticides and a desire to provide adequate controls, either voluntary or mandatory, to ensure that the use of pesticides does not affect public health.

APPENDIX

The following notes summarise information on the extent to which existing legislation in the UK covers (or could be used to control) various aspects of the supply and use of pesticides. (They should not be regarded as legal interpretations of powers available under the Acts.)

Import, Export and Customs Powers (Defence) Act, 1939

The powers derived from this Act and exercised under the Import of Goods (Control) Order, 1954 are very wide, but are mainly designed to control imports for what might be described as broad economic purposes. They are used for other purposes such as the control of the importation of firearms and explosives and the conservation of species. It might be inappropriate to take steps legally to control imports of pesticides to plug possible loopholes in a voluntary scheme.

Pharmacy and Poisons Act, 1933

The Pharmacy and Poisons Act does not provide for absolute prohibitions on sale of poisons nor for control over importation. Distributors cannot therefore be prevented from importing a particular chemical that has not been cleared under the Pesticides Safety Precautions Scheme. However, distribution might be considerably hampered if the chemical concerned were a listed poison in respect of which especially strict limitations on sale or supply have been imposed (for example, strychnine and monofluoroacetic acid).

Farm and Garden Chemicals Act, 1967

Regulations made under the first part of this Act came into operation on 1 May 1973. They schedule some 290 chemicals and require the containers of insecticides, weedkillers and other pesticides sold for use on farms and in gardens to be labelled clearly with the names of any of the scheduled chemicals they contain; also, if a pesticide is sold without a container or in one provided by the purchaser, such a label will have to accompany it. Regulations covering the second part of the Act, requiring any such label to bear a mark, symbol or colour to indicate the extent of any hazard which the product offers to man or other forms of life and to bear prescribed words of explanation or warning, are envisaged in the not too distant future.

Trade Descriptions Act, 1968

The broad purpose of this Act is to ensure that the buyer of goods is not deceived by the seller as to the nature of the goods—in short to protect him from economic harm. Section 8 powers enable the Department of Trade and Industry to impose a requirement that any goods shall be marked with or accompanied by any information which they regard as necessary or expedient in the interest of persons to whom the goods are to be supplied. Broadly similar powers exist under Section 9 which require the same information to be given in advertisements for goods. In general, orders made under these sections would apply across the board to both home-produced and imported goods of the categories to which they related. These order-making powers are regarded as being applicable to factual matters in relation to the goods such as their composition or origin which the consumer may need to know: but they are unlikely to be available to impose warnings about possible dangers in the use of the goods. The Consumer Protection Act is generally regarded as the proper medium for the imposition of controls over dangerous articles.

In practice there are definite legal limits in using the Act to require goods to carry warnings of dangers arising from their use; however, the Act could conceivably be used to impose requirements regarded as conducive to greater safety in the use of pesticides.

Consumer Protection Act, 1961

This Act empowers the Home Secretary to make regulations prescribing safety requirements relating to any class of consumer goods if, in his opinion, such action is necessary to prevent or reduce the risk of death or 'personal injury' (defined as including disease or disability). The regulations may relate to such matters as composition or contents, design, construction, finish or packing of goods and may require consumer goods to be marked with or accompanied by any prescribed warning or instructions. It is an offence under the Act to sell or offer for sale any goods which do not comply with such regulations. Manufacturers, importers and wholesalers as well as retailers can be prosecuted for such an offence.

Thus if a given pesticide presented a serious risk to the individual user, appropriate regulations under the Act could no doubt be considered. But it should be noted that this Act was designed to deal with risks in manufactured articles offered for sale to the public and has hitherto only been used for this purpose. Whether regulations could be applied to chemical substances as such is a question not entirely free from doubt.

Protection of Animals Acts, 1911 and 1927

These Acts make it an offence knowingly to put or place (or cause or procure any person to put or place, or knowingly to be a party to the putting or placing) in or upon any land or building any poison, or any fluid or edible matter (not being sown seed or grain) which has been rendered poisonous.

It is a defence that the poison was placed by the accused for the purpose of destroying insects, rats, mice or other small ground vermin, in the interests of public health, agriculture or the preservation of other animals, domestic or wild, or for the purpose of manuring the land, and that he took all reasonable precautions to prevent injury to domestic animals and wild birds.

Animals (Cruel Poisons) Act, 1962

This Act enables the Home Secretary to prohibit or restrict the use of any named poison for the purpose of destroying mammals, if he is satisfied that the poison cannot be used without causing undue suffering and that other suitable and adequate methods of destruction exist. Regulations made under the Act prohibit the use of phosphorus and red squill, and strychnine (except for killing moles).

Protection of Birds Acts, 1954 and 1967

Under these Acts it is an offence to use any poisoned, poisonous or stupefying substance for the purpose of killing any wild bird, or to place any such substance so as to be calculated to cause injury to any wild bird. Licences may, however, be granted to use poisoned or stupefying substances for the purpose of killing certain named pest birds, subject to suitable precautions being taken to prevent harm to protected species.

Food and Drugs Act, 1955

The Act makes it an offence to add a substance to food (including harvested crops), or to subject food to a process or treatment, if the food is thereby made injurious to health and if it is to be sold for human consumption. The Act also makes it an offence to sell food which is unfit for human consumption or to sell to the prejudice of the purchaser food which is not of the *nature, substance or quality demanded.* Regulations made under the Act control the presence in food of specific substances, for example arsenic, lead, fluorine, and also preservatives (some of which may also have pesticidal uses).

Enforcement may be at any stage in the distributive chain and so long as food can still be identified, responsibility for an offence can be passed back down the distributive chain. Normally, loss of the food's identity would mean that proceedings could only be taken successfully against a farmer who misused pesticides on his crops if he sold direct to the consumer or to a retail outlet.

Agriculture (Poisonous Substances) Act, 1952 and Regulations

The purpose of these regulations is to protect farm workers from poisoning by the more hazardous compounds used in agriculture by ensuring, among other things, that they are supplied with and use certain protective clothing when performing scheduled operations with specified pesticides. Other obligations concern training, age limit of persons employed on scheduled operations, keeping of records, limits of work-time on scheduled operations and washing facilities. The leaflet APS/1 summarises the provisions of these regulations.

Rivers (Prevention of Pollution) Acts, 1951-1961

This legislation makes it an offence to cause poisonous, noxious or polluting matter to enter a 'stream' (which includes any river, stream or watercourse or inland water discharging into a stream). It is possible to prosecute after an offence has been committed but there is no specific legislation to require care in using or handling pesticides to avoid as far as possible accidental pollution which might contaminate sources of water supply.

Salmon and Freshwater Fisheries Acts, 1923 and 1965

These Acts make it an offence, with certain exceptions, to put any liquid or solid matter into fishing waters (i.e. all waters in which fish exist, including ponds and lakes) with the result that the waters are rendered poisonous or injurious to fish. Amending legislation is in prospect which would, among other things, increase the penalties.

REFERENCES

Council of Europe (1969). *Agricultural Pesticides*, 2nd edn. Strasbourg; Council of Europe

Ministry of Agriculture, Fisheries and Food (1972). *Approved Products for Farmers and Growers*, (Annual list). London; HMSO

Ministry of Agriculture, Fisheries and Food. *Chemical Compounds used in Agriculture and Food Storage, Recommendations for Safe Use in the United Kingdom*, dossier with constant additions and revisions. London; HMSO

EEC (1967). *Draft Council Directive fixing the Maximum Residues Content for Pesticides on and in Fruits and Vegetables*, Brussels; EEC

EEC (1971). *Draft Council Directive fixing the Maximum Contents for the Residues of Pesticides on and in Raw Cereals*, Brussels; EEC

FAO (1970). *A Model Scheme for the Establishment of National Organisations for the Official Control of Pesticides*, Rome; FAO
MAFF (1969). The Safe Use of Poisonous Chemicals on the Farm', Revised July 1969. London; Ministry of Agriculture, Fisheries and Food
Panel on Residues of Pesticides in Foodstuffs (1966–67). 'Pesticide residues in the total diet in England and Wales, Parts I–IV', *J. Sci. Fd. Agric.* Vol. 20, 242, 245: *Pestic. Sci.* Vol. 1, 10, 99
Panel on Residues of Pesticides in Foodstuffs (1966–69). 'Pesticide residues in Foodstuffs in Great Britain, Parts I–XII', *J. Sci. Fd. Agric.*
Panel on Residues of Pesticides in Foodstuffs (1970–71). 'Pesticide residues in Foodstuffs in Great Britain, Parts XIII–XV', *Pestic. Sci.*
Pesticides Safety Precautions Scheme agreed between Government Departments and Industry (1971). Ministry of Agriculture, Fisheries and Food, London.
Report on Safety, Health, Welfare and Wages in Agriculture. (Annual reports). London; HMSO
Review of the Present Safety Arrangements for the Use of Toxic Chemicals in Agriculture and Food Storage (1967). London; HMSO
Royal Commission on Environmental Pollution. First Report. Feb. 1971. London; HMSO

FURTHER READING

Bates, J. A. R. (1968). 'Reflections on regulations—certain aspects in the official control of pesticides in various countries', *Chem. and Ind.* 1968, 1324–32
The Collection of Residues Data (1969). London; HMSO
Control of Pesticides—a survey of existing legislation (1970). Geneva; WHO
Further Review of Certain Persistent Organochlorine Pesticides Used in Great Britain (1969). London; HMSO
Pesticides in the modern world (1972). A symposium prepared by members of the Co-operative programme of Agro-allied Industries with FAO and other United Nations Organisations. (Obtainable from British Agrochemicals Association, London)

DISCUSSION

Knights: You have been talking mainly about pesticides—does the Pesticides Safety Precautions Scheme not cover other agrochemicals such as, for example, antibiotics and growth promoters.

Bates: The definition of 'pesticide' in the Scheme includes *all* agrochemicals bar fertilisers and veterinary medicines. The latter are covered by the Veterinary Medicines Act but a few veterinary chemicals not directly administered to animals (for example barn sprays) are still covered by the Scheme.

Knights: You made a special point about labelling of products but is the ordinary farmer given any more information about the actual toxicity of the material he is using?

Bates: One of the aims of the scheme is to give advice on the safe use of a product. It does this through the Official Recommendations which are available to everyone but go mainly to Public Health Officers, Libraries, and so on. The main way, however, is via the product labels—each one is carefully scrutinised and agreed with the company. Industries' attitude is that they have subscribed to the Scheme since its inception and it is part of their way of life, so to speak, and they include the official recommendations on their labels. The label is the main means of telling the farmer or gardener how to use the product efficiently and safely.

I believe that the MAFF could improve the public relations aspects of the Safety Scheme. Safety measures are not very newsworthy and thus get little publicity. As many people have said in this Symposium, agrochemicals tend to get only unfavourable publicity and there is plenty of scope for favourable publicity by both government departments and industry in this respect.

Nine

The economics of the use of chemicals in agriculture

K. E. Hunt
Institute of Agricultural Economics, Oxford

SUMMARY

Agricultural practices vary throughout the world and these interact so closely with the agricultural policies of national governments that discussion cannot be restricted to any one national situation. Economic aspects of the use of agrochemicals affecting farmers in both the developed and underdeveloped countries are therefore considered. Farm-level decisions are also greatly affected by national and world economic situations, as are supply and prices to the consumer, and such external factors are reviewed, especially with regard to differences in availability of labour and capital and in farm price support systems in different countries. Possible alternatives to using presently available agrochemicals are discussed with special reference to the interaction of these alternatives with policy measures affecting home agriculture and the food potentially available for underdeveloped countries.

Great problems are experienced in making quantitative estimates of the benefits and costs to the farmer and consumer of the use of agrochemicals, especially in relation to possible damage to the environment. A plea is made that governments should recognise the problems involved and the need to make such assessments, and that they should provide the investment in human skills and organisation necessary to improve analyses, the machinery for full and rational use of evidence and the subsequent decision-making and enforcement procedures.

INTRODUCTION

Between the diversity of chemicals used, the diversity of agricultural systems in different parts of the world, and the several points at which relevant decisions concerning the economics of chemical use are taken, this subject is potentially a very broad one. In this chapter, the chemicals considered will be fertilisers, insecticides, fungicides and herbicides. Their use represents a sophisticated form of technology, the adoption of which will be influenced by the ability of farmer-operators to learn techniques which differ markedly from traditional practice and their ability to apply these in the prevailing economic conditions. Agricultural industries in the various countries of the world, together with the agricultural policies pursued by national governments, so closely interact with each other that it is unreasonable to confine this economic discussion to any one national situation. Consequently there is need to take into account, on the one hand, agriculture in industrialised countries and, on the other, agriculture in the developing countries.

Agriculture in the former includes general mixed farming, specialised livestock production (whether of broilers, cattle in feed lots or other intensive systems), fruit and vegetables grown on open land and under glass and production based on hill and range land. In different situations, producers may range in size from tiny businesses contributing only part of the income necessary to maintain a single family up to very large complexes of business activities. Moreover, in some instances the agricultural enterprises are wholly independent, being linked to suppliers of their raw materials and purchasers of their products only through the indirect means of traditional marketing practices. However, in other instances, varieties of contracts and other forms of vertical integration link very closely enterprises operating at several levels in the production and distribution chain. Similarly, various kinds of horizontal integration—co-operatives for production, for distribution and for the purchasing of requisites—link businesses of similar kinds operating at the same level. Generally speaking, in the industrialised countries, agricultural labour is relatively scarce and capital, though not as freely available as producers might wish, is nevertheless not often a major limiting factor for production systems that are reasonably economic. The education, general and technical, of farmers under these diverse conditions naturally varies a good deal, but typically they are literate and there are available trained advisory officers in government or privately organised advisory services who can assist farmers with technical and economic decisions.

Agriculture in developing countries is almost equally diverse. In some, land-use may take the form of a system of shifting cultivation under which forest or grassland is cleared, crops planted for a few years, after which the land reverts to forest or grassland. In other cases, intensive systems may be typical of the practice, for example vegetable production around large towns or irrigated rice lands. Generally labour is in ample supply though locally it may be short—for example where there is a substantial migration of male workers to mining or other paid employment. Capital may in fact be available on a larger scale than is customarily believed but its mobilisation for investment purposes is usually difficult. Though advisory services for particular purposes may be available, the intensity of coverage is usually much less than it is in developed countries. Types of vertical integration may be found in developing countries as well as in developed ones, for example where small farmers are mobilised to produce fruits or vegetables for processing. In these cases technical guidance—though possibly not economic guidance to the same degree—is usually made available by the contractors to co-operating producers. There are also examples of dual economies where, in addition to small scale peasant production, large scale, centrally organised, plantation production of such products as, for example, tea, palm oil and sisal, is carried on. Though cases are found of local financing of these enterprises, they are typically associated with expatriate capital and management, and the profits of the enterprise may be, to a large extent, exported. Technical and business advice is not likely to be a limiting factor for these enterprises.

Situations where decisions have to be made concerning the economics of the use of chemicals in agriculture can be identified at several levels. A review of economic aspects would need to consider:

1. Economics of chemical use for the individual farmer.
2. The fact that any pollution they cause represents a cost not included in the farmer's decision data.
3. A means of assessing this 'external' cost.
4. The fact that many national agricultural industries are supported by subsidies or protective tariffs and the use made of chemicals must be considered against this background.
5. The fact that rich and poor countries differ greatly in the adequacy of food supplies available to them with the possibility that special measures to achieve greater equality may affect the economic incentive to use chemicals.

For the present purpose the decisions taken outside the farm seem the more important because they are less commonly discussed. Consequently this discussion proceeds first to a brief outline of the economic issues at the farm level and then to present some of the non-farm issues.

FARM SITUATION

In many instances it is clearly highly profitable to apply chemicals in agricultural practice, hence the increase in their use. For example, Indian studies have shown an increase in revenue equal to five times the cost of an application of nitrogen and a study in the Philippines (see *Table 9.1*) gave an increment in revenue over twelve times the increment in nitrogen cost when growing rice (for other studies on the economics of fertiliser use, see CBAE, 1971 and 1972). Instances of high gains also can be recorded for herbicides and pesticides.

Table 9.1 EXAMPLES OF CHANGE IN COSTS AND RETURNS ASSOCIATED WITH INCREASES OF 30 kg NITROGEN PER HECTARE APPLIED TO RICE IN THE PHILIPPINES (*Courtesy: International Rice Research Institute, 1967*)

N *applied* (kg/ha)	*Added return*	*Added cost* (pesos per hectare)	*Added return/ added cost*
0	0	0	0
30	444	35.4	12.6
60	311	35.4	8.8
90	177	35.4	5.0
120	43	35.4	1.2

Though the profitability of chemical application, leaving aside any external costs, may be quite clear at lower levels of application, at some stage, as input is increased, the problem becomes more difficult. For example, as *Table 9.2* shows, the uncertainty of the response may increase appreciably with increase in nitrogen fertiliser application. There may be differences too in the response between varieties and between seasons. Since it seems likely that the increased variability with increased nitrogen application is connected with the adequacy of the water supply to the growing plant, it is probable that the shape of these curves will be influenced by the presence or absence of irrigation facilities. Many published figures represent the results of studies of a single season and conceal the year to year variation and the economic uncertainty it represents. There may be cases, too, where the expression of the economics of chemical use is difficult to quantify. For example an enquiry as to the use of herbicides in the UK (Evans, 1967) showed only a three-in-seven chance of increasing yield. A formal assessment of the place of this chemical in the situation would have to take account of the problems of getting labour to operate systems which are capable of dispensing with chemical weed control techniques, but which are more labour intensive than the present ones. It is usually very difficult to get firm data for the cost of producing specific qualities of a crop by methods which dispense with the use of chemicals. This would require the use of quite different production systems, and examples indicating the probable countrywide behaviour of such systems are rarely available for recording when prevailing systems are predominantly based on chemicals.

Table 9.2 EXPECTED NET REVENUE AND VARIANCES OF NET REVENUES ASSOCIATED WITH USE OF DIFFERENT NITROGEN LEVELS IN NORFOLK SOILS OF THE NORTH CAROLINA COASTAL PLAIN, USA (*After: Tollini and Seagraves, 1970*)

Nitrogen applied (lb/acre)	*Expected net revenue* ($/acre)	*Variance of net revenue due to the weather* ($/acre)
0	36.41	2.40
50	68.53	68.77
60	70.04	79.86
70	71.20	90.79
80	72.07	101.60
90	72.70	112.29
100	73.14	122.91
110*	73.41	133.45
120	73.54	143.92
125*	73.63 (max.)	149.14
130	73.54	154.34
140	73.43	164.71
150	73.22	175.03
200	70.96	226.11

* To maximise expected net revenue the farmer should apply 125 lb/acre but by reducing his input to 110 lb he would cut his revenue by only 0.3 per cent to $73.41, but he would reduce the variance by 10 per cent

Over all, a producer's choice must depend, in part at least, on the following considerations.

1. Factors which affect per unit cost of alternative inputs, for example labour.
2. Knowledge which can permit the farmer to vary the input/output relationships.
3. Knowledge which affects the assurance with which a farmer can assess the input/output relationship which will face him. This in turn has two elements, the normal shape of the curve to which he has been accustomed and the shape pertaining to a coming situation when new price regimes will be operating in consequence of new supply situations arising from general application of new technologies.

DECISION PROBLEMS LYING OUTSIDE THE FARM STAGE

A review of economic aspects of the use of chemicals would need to take some account of at least the following considerations which influence the economic situation in which the farm-level decisions have to be taken.

Factors affecting costs of chemical inputs

Obviously this is a complicated area and its details cannot be pursued here. Nonetheless, the price at which farmers can buy chemical inputs, and the sophistication of those available, must be expected to depend on the economic environment of the chemical manufacturing and distributing industries in the case of products produced within the country under consideration. Where they are imported it will clearly depend on the import conditions. A review of the technological economics of the manufacturing side will presumably include the problems of integration of the particular production line concerned into the over-all programme of the manufacturing firm and planning and conduct of research, the screening of routinely synthesised products for useful herbicidal or other activity, and the uncertainty attaching to the obsolescence of the product arising from competition from other producers' lines, or the development by the pest of biological resistance to the active constituent. There are also aspects of the width of the market and the extent to which the product fulfils a specific need not covered by extension of the application of other products which have a different spectrum of application. The situation will also be affected by the availability of alternatives to the use of chemicals. Some of these may be simply the acceptance of a lower production of the farm product, a situation discussed in the next section. Often the alternative to using chemicals would be a major change in the farming system in order to bring about the increment in output or control of pathogens, required by husbandry techniques. Though the manufacturers of agrochemicals no doubt carry out a certain amount of comparative research into non-chemical control as part of the overall testing of the possibilities of their products, it is not to be expected that they will actively develop procedures for controls for pests and weeds which require no chemical sale product. Consequently, the considerable body of resources for research and testing represented by the chemical industry is not fully available for development of alternative methods and the problem arises of how such research into alternative systems could be carried out.

In many instances, if pests and weeds are to be controlled other than by chemical means, the use of more labour—and possibly more capital—will be required. In assessing the economic consequences of varying the mix of chemical and non-chemical methods, even where the physical input/output relationships are known, there is clearly a problem of valuing labour correctly. Where it is hired under conditions of reasonably full employment, its market price probably gives a reasonably realistic assessment of the proper value to be entered in the calculation. However, where family labour is involved—as it usually is for many small producers who are on the edge of the market economy—or where labour is hired but where there is a labour surplus, the correct valuation is very much less easy. In many developing countries there is an additional complication in that the choice open to policy makers is between a situation where foreign exchange is used for the import of chemicals manufactured abroad and the use of relatively costless local labour. This applies particularly to chemicals for pest, disease, and weed control since a number of developing countries are now able to manufacture their own fertilisers. It should, perhaps, be noted that it is not always clear whether the price at which agrochemicals move in international trade reflects a fully competitive situation.

There are, however, in addition to these matters concerned with intra-farm decisions and the pricing of chemical inputs, two broader issues. One of these concerns the price–supply–demand environment for farm products in which the agricultural producer makes his decisions. Obviously the price of agrochemicals is intimately involved in this situation. The second problem is largely a non-market one; namely it relates to 'externality problems'. These include all the environmental and other matters where the action of entrepreneurs making their private decisions brings costs or benefits to others and no mechanism is available for providing recompenses or benefits to offset them.

Turning first to the product supply situation, it must be recognised that almost the whole of world agriculture is an artifact of the decisions of governments expressed in their policies towards agricultural industries. Thus in Britain for more than forty years there have been various forms of protection of prices received by farmers, now largely incorporated in the agricultural price support system developed under the 1947 and 1957 Agriculture Acts. Under these arrangements the price to farmers for their product is separated from the price to consumers by an element of subsidy—a transfer payment from the taxation receipts of central government, i.e. the taxpayer, to supplement what the market provides to the farmer. The transfer from consumers to producers of farm products is arranged under the Common Agricultural Policy of the EEC by a different means, namely by tariff protection at the frontiers of the community, resulting in raised prices for the product and its substitutes throughout the community. The prospect is for Britain to move to a similar type of protective arrangement in the next few years. It might be noted that in the UK at present the transfer payment comes from the better off sections of the community, i.e. the taxpayers, whereas in the future it will come from the whole body of consumers. The USA is operating other arrangements and the various developing countries have intervened in the market for farm products, sometimes to supplement producers' income, sometimes to tax it. Obviously the agricultural industries of the centrally planned economies reflect government policies towards them.

Where farm prices are supported, if an agrochemical is introduced and current prices for the chemical and the farm product concerned make it attractive to producers, they will make increasing use of it, resulting, in most instances, in increased product output. The question which is then posed is should the support price be left as it is, or be reduced? If the support price is maintained, farmers will gain the full benefit of the new technology. The amount of funds transferred from the taxpayer to farmers under a support system must be expected to increase, both because market prices would otherwise fall and consequently the margin to be made good by income supporting action will be increased, and also because the quantity which attracts support action will be increased. If the support system is of a tariff protection kind, then surpluses are likely to develop since more will be produced than will move into consumption at going prices for the products. Some means of removing these from the market and disposing of the surplus will have to be found. If this is directed to particular sections of the population under some form of low income relief scheme, as for example has occurred in the USA, some sectors of the consumer–taxpayer group will benefit and others will not. If supplies are disposed of overseas, then the loss to consumer–taxpayer will be complete. (It may be that, as a body, consumer–taxpayers might be prepared to contribute to the relief of overseas populations who are suffering distress for lack of food by sending them food. Even if they are, it is a valid argument that they should have the opportunity to decide on this explicitly and not have the contribution imposed on them by the accident of administration of a farmer–income support scheme.)

If, in contrast to the retention of prices at the pre-innovation level, they are reduced, then part or all of the gain will pass to the consumer–taxpayer. Though the issue has not arisen explicitly in connection with chemicals in agriculture, the issue of the distribution of returns from new technology is recognised in the procedure operated under the terms of the agricultural protection system in the UK. As a working basis, some £25 million is assumed to be a fair contribution by farmers to the consumer–taxpayer out of the gains from improved technology and efficiency each year. Naturally, farmers protest strongly in public about this adjustment, arguing that it requires initiative and effort to introduce new technologies and it is unfair that they should be robbed of the benefit from them by this device. On the other hand, it is relevant to note that without some sort of guarantee system they are likely to lose completely the benefit of technological change.

Bearing these considerations in mind, one might now look at what is involved if the use of chemicals in agriculture is discouraged. Among the possibilities are the following.

1. If production is to be maintained:
   (a) Use may be made of a less noxious but presumably more costly chemical in place of the one to be phased out. It is presumably a more costly one since otherwise it would be preferred by producers to the one under discussion. Such extra costs may, however, be complicated by the differences in the characteristics of the two chemicals. Thus one, the more costly, while performing the same service as the other, may provide either greater certainty or a wider spectrum of action neither of which, in the particular situation, is considered of practical relevance.
   (b) More labour may be used either, for example, for weeding or for the individual removal of pests or diseased leaves.
   (c) More land may be brought into cultivation if free land is available.
2. Production might be allowed to fall. An illustration of one such calculation is given in *Table 9.3*.

In either case there is the problem of the price policy which has to be faced, whether to leave farm prices unchanged, or to alter them wholly or partially to offset the change in costs incurred as a result of the proposed change.

The above considerations apply in the shorter term. In the long term, it is open to society to so arrange investment in research that new non-chemical technology may be introduced which would represent a different, presumably improved, relationship between non-chemical inputs and outputs.

Considerations applicable to developing countries

In principle, most of the considerations outlined above apply also to the developing countries, though their practical significance may

Table 9.3 ESTIMATES OF THE COSTS ($ MILLION) ASSOCIATED WITH RESTRICTING THE USE OF PHENOXY HERBICIDES ON DIFFERENT CROPS IN THE USA IN 1969 (*After USDA, 1970*)

		Additional costs			
Crop	*Reduced costs of materials and application*	*Use of substitute herbicides and application*	*Additional cultural practices*	*Production on additional acres*	*Net additional costs*
Corn	−37.0	122.5	21.2	—	106.7
Wheat	−21.9	15.3	12.1	45.0	50.5
Other small grain	−14.6	10.9	9.1	23.1	28.5
Sorghum	− 5.6	14.5	2.4	—	11.3
Rice	− 0.4	—	6.4*	1.6	7.6
Other crops	− 5.4	—	—	21.3	15.9
Pasture	−10.4	—	43.3	—	32.9
Range	− 7.2	—	43.1	—	35.9
All crops	−102.5	163.2	137.6	91.0	289.3

* Includes $2.2 million for lower income from loss in quality

be different. A considerable amount of attention has been devoted to the study of factors affecting the uptake of innovations by producers. There is the straightforward problem of making the technology available to them, and the problems of mobilising the necessary capital for the purchase of chemicals and other non-farm inputs and the provision of such equipment as sprayers. There is also, however, the problem of taking due account of the uncertainty elements which must exist in any situation of this kind. When using traditional methods, producers have gained from experience a fair idea of the risks attached to them, but when using new technology, particularly practices which give useful but not massive increases in return, the producer has great difficulty in making a fair assessment of the uncertainty involved. This problem is enhanced where there is an increase in the degree of uncertainty with the level of inputs of the chemical concerned.

At the policy level, apart from matters concerned with schemes to provide credit and to encourage uptake of new technologies, there is the question of the source of inputs. It may be fairly straightforward to assess whether it is worth while importing fertilisers to produce an export crop; it may be very much less straightforward to judge the usefulness of importing them rather than to use available manpower—at the present unemployed or only partially employed. The manpower could be used to cultivate land currently unused or to make possible double cropping with sale crops for internal use, or double cropping with a soil enriching catch crop, in place of simpler systems formerly used.

In addition to the problems discussed above, there is a group of quite a different nature. Notwithstanding the expansion in output resulting from the 'Green Revolution', the world food picture which seems to be inexorably emerging is one in which food production will tend to exceed demand in the developed countries and to be below demand in the developing countries. Clearly, different assumptions about the spread of new technology and of price policies and the like will vary the particular picture but the general indications seem common to various assessments. (Gittinger, 1970; OECD 1971). This poses a difficult problem when a decision has to be taken in developed countries as to whether a high yielding technique should be abandoned because of the environmental harm it causes with a resulting temporary or permanent fall in food production. If the choice was as clear cut as this, then it might be expected that a reasonably clear decision could be reached. There are many sincere students of the problems of development, and agricultural development in particular, who have strong reservations regarding the usefulness to the recipient countries, even in the short term, of donations or concessional provision of food products from the developed countries. Many developing countries, furthermore, look on this with distaste for political reasons, the continuance of colonial situations and so on.

EXTERNALITIES

It is not appropriate here to discuss the technical aspects of the effects of the use of chemicals in agriculture on those outside the agricultural sector, including such issues as the build-up of nitrogen in natural bodies of water, of DDT in food chains, and the loss of health and of human enjoyment from environmental deterioration. These have been the subject of the earlier chapters. On the whole it appears that the view has been taken that the risks under these several headings are rather small and well under control, at least in the UK. But not everyone would perhaps be equally assured of this or, if they were, would be free of anxiety about the future situation or the state of affairs in other countries. These questions therefore have to be faced in any economic assessment of the use of chemicals in agriculture. It is unrealistic to think that one can isolate any national agricultural industry indefinitely from the influence of conditions in agricultural industries of other countries, notwithstanding the efforts of governments to insulate one from the other.

There are at least three broad problems which should be borne in mind under these headings.

1. Overcoming the difficulty of getting a factual assessment of the situation on pollution and the like, accurate enough for working purposes, across all environments in respect of the many relevant aspects.
2. Forming a view—if not making a measurement—of the effect of changes in the input of chemicals into agriculture on the detrimental effects coming under this heading of 'externalities'. For example, by how much does one need to reduce the use of nitrogen in order to maintain lakeland environments in an acceptable condition?
3. If at all possible, finding some common terms for valuing these effects so that at least the outcome of different measures can be compared, even if the measurements cannot be taken into a more or less orthodox comparison of costs and benefit in cash.

VALUATION

A great deal of ingenuity has been introduced into the provision of prices of inputs and outputs for use in calculations in this field. (For a manual of cost–benefit analysis, see Little and Mirrlees, 1969.) However, there comes a stage, though different people might differ in their view on just where it is reached, when the constructs are too fragile to be useful. A very great deal of work remains to be done in this field, but it seems at least possible that the most useful lines of work might include the following.

1. To avoid trying to put a measure in £ sterling on, for example, the value of one acre of mountain land whose animal or plant species complement has been reduced in consequence of chemical action spilled over from farming, but rather to try and see what we would need to forego in agricultural production in order, say, to save one acre of such land.
2. Given some such calculations as these, it should be possible to be broadly consistent in our activities in different environment-improving or life-saving directions. For example, rationally we should be prepared to layout about the same amount of money to reduce motor accidents by one per year as deaths by toxic chemicals by one per year or to save one acre of attractive ground from coal tips as from cement ash.

Though one might hesitate to incorporate at the individual 'firm' decision level the cost element derived from externalities, there are various policy decisions where it would be appropriate to give weight to it. Certain experience has been built up for comparative purposes within the sphere of problems of external costs which might be extended to situations involving farm chemicals. For example in respect of injury to human beings, the cost of diagnosing and treating, the cost of lost output and of tending the handicapped have been regarded as reasonably appropriate measures of cost of injury in certain cases. For loss of life, the cost of rearing a child or the cost of saving life by the cheapest alternative means might provide some basis for a check on an emotional approach to comparison. If assessments along these lines prove inoperative, it may be helpful to search for maximum levels of chemical use which would have no significant deleterious effect. (For a review of these problems, see Headley and Lewis, 1967).

In reading much of the literature it would seem that even without success in appraising the valuation or formal incorporation of external costs and benefits, a useful advance in this field would be achieved if a wider range of people became accustomed to discussing the subject and appraising damage and benefits with a knowledge that there are costs and benefits applicable to all the possible lines of action in a given situation, that the weight is not all on one side. If this, and the fact that a choice has to be made after weighing up the costs and benefits of the several alternative lines of action as rationally as possible, is thoroughly realised, some advance will be achieved.

CONCLUSIONS

It seems clear that some elements in the economics of chemical use in agriculture are readily quantifiable. The statistical material is by no means fully adequate for the purpose, especially in relation to uncertainties of outcome, but a reasonable review can be made. For other elements, however, where the market does not operate, assessment is much less easy. Cost–benefit analysis can help in this respect, but at a certain stage the ingenuity required to provide a valuation for the various external effects of the use of chemicals appears unrewarding, if not indeed self-defeating. These complexities clearly have to be lived with and the tools of the accountant and the economist do not seem fully applicable to their solution. Various techniques, some touched on in this chapter, have a contribution to make, but they have their limitations. It seems highly important that one should recognise the problems which this presents and an important conclusion seems to be that there is an urgent and grave need to provide the necessary investment in human skills and in organisation to improve the handling of decisions in this field.

Governments are clearly involved in many of the decisions concerned here, either directly or indirectly, through the policies which influence the decision environments of the individual firms, but governments are not traditionally well-organised to co-ordinate matters of such diverse description. Co-ordination tends to be much better within departments than it is across departmental boundaries. Clearly quantitative methods have to be used as far as they will go, methods of experimental science as far as they will go, aesthetic judgments where these are applicable, political judgments where they in their turn may be applicable. But by the nature of humanity, it is impossible to get expertise in all these fields contained within one individual. Consequently it will be necessary to have working groups, not only of a consultative nature but of an operational and administrative nature, which between them provide such expertise. However, if such groups merely consist of a number of experts each specialised in their own field, it seems highly likely that they will either achieve no action or a sequence of inconsistent actions determined by the chance course of discussion or the force of personality of individuals. It seems critically important, and this might be the most important outcome of this review, to have someone serving such a group who knows enough of all facets, particularly the economic, the technical, and the social, to be in a position to force the individual specialists to table adequately prepared cases. The physical planners concerned with general resource planning are moving towards this situation. There seems to be an urgent need for a class of individuals trained in biological and social science who are capable of ensuring that the several considerations involved in decisions in which the economics of the use of chemicals in agriculture are part, are fully, evenly and relevantly displayed in preparation for as rational a decision as it may be humanly possible to achieve.

REFERENCES

CBAE (1971). *Annotated Bibliography No. 4.* Commonwealth Bureau of Agricultural Economics, Farnham Royal, Bucks, England

CBAE (1972). *Annotated Bibliography No. 4, Supplement A.* Commonwealth Bureau of Agricultural Economics, Farnham Royal, Bucks, England

Evans, S. A. (1967). *Proceedings of Joint Conference organised by National Farmers Union and Royal Society of England*, 12th January 1967 (mimeo)

Gittinger, P. (1970). *North American Agriculture in a New World*, National Planning Association, Washington DC

Headley, J. C. and Lewis, J. N. (1967). 'The pesticide problem: an economic approach to public policy', in *Resources for the Future*, John Hopkins Press

International Rice Research Institute (1967). *Annual Report 1967*, 246. Los Banos, Laguna, Philippines

Little, I. M. D. and Mirrlees, J. A. (1969). 'Social cost benefit analysis', in *Manual*

of Industrial Project Analysis in Developing Countries. Vol. II. OECD Development Centre, Paris
OECD (1971). *Food Problems of Developing Countries*, OECD, Paris
Tollini, H. and Seagraves, J. A. (1970). *Actual and optimal use of fertiliser. The case of nitrogen on corn in eastern North Carolina.* Economic Research Report No. 14, 13. Dept of Economics, North Carolina State University
USDA (1970). *Restricting the use of Phenoxy Herbicides. Costs to Farmers.* Agricultural Economic Report No. 194. USDA Economic Research Service and Agricultural Research Service

DISCUSSION

Reay: What do you feel about the fact that the use of pesticides is actually contributing to the world population problem by reducing starvation?

Hunt: This is perhaps a moral philosopher's rather than an economist's problem to judge, but my feeling is that extra mouths to feed result certainly, but aren't there also extra pairs of hands then available to cultivate the land and utilise more fully available technology? Some economists say the population/food imbalance is a transient phenomenon but I don't myself feel that optimistic, and would like to see some form of population control instituted.

Reay: I feel that one of the points that hasn't been made in this Symposium is that one cannot really separate population control and pollution.

Griffin: Are you saying, Mr. Hunt, that in comparing labour intensive societies and capital investment societies, the developing countries have an asset in terms of labour availability? Is this an alternative to capital investment growth for such countries?

Hunt: To a degree, yes, I think this is an asset. Capital elements are necessary to produce the fertilisers and pesticides which are an essential adjunct to the Green Revolution. But in many countries, there doesn't seem to be anything but labour in abundance. How you organise this is one of the biggest problems facing us.

General discussion

Knights: This is now an appropriate point at which to summarise the general conclusions that appear to have been reached in this symposium. From this, many questions will no doubt arise which can be usefully discussed.

One fundamental point that Mr. Mason put in his opening talk and which has been emphasised since, is that with increasing populations, industrialisation and urbanisation, agricultural productivity must also expand, despite reductions of suitable acreage. This, it has been said, can only be achieved through the use of agrochemicals.

Other speakers have discussed the needs for various categories of such materials, their modes of action and their possible pollutant side effects. The suggestion has been made several times, however, that the pollution problem attributable to agrochemicals is very often over-exaggerated and that alarmist views receive undue publicity.

It appears then that the consensus of opinion from this Symposium is that the use of agrochemicals is essential and the pollution problem must be discussed with this premise in mind. This view is, however, tempered by great awareness by scientists of the possible pollution side effects, and the extensive and intensive research into the problem by industry, the ministries, universities and polytechnics has been outlined and emphasised by many speakers.

But some questions arise from this—will such research actually solve the problems or minimise them to an 'acceptable' level? What is an 'acceptable' level? Are we perhaps being too complacent and is more drastic action required? Is stricter legislative control needed? Also, what are the costs involved in using or not using agrochemicals and who should bear the costs? And what of the world-wide situation?

Another important question we must ask ourselves—how can the more moderate views such as those expressed in this Symposium be broadcast widely and what is the future role for education and information dispersal? Finally, we have not considered ethical and aesthetic arguments in any depth and our discussion can now embrace these too.

Bates: It seems to me that the first question we should ask is what evidence is there that agrochemicals produce any measurable pollution in the UK? For example, MAFF has looked at the various aspects of mercury pollution. Newspapers stated that mercury was used primarily as a fungicide but in fact, on examination, it was found that the amounts used in agriculture are small and present very little of a problem. In relation to the amounts of mercury used in cereal seed dressings, one could apply these rates to arable fields for a hundred years with only a very slight change from the natural background levels in the soil.

Reay: How in fact does one define 'pollution'?

Bagnall: *Bayer Agrochemicals Ltd.* Is it not any material found out of its normal place that has an adverse effect on living organisms or non-target organisms?

Knights: Surely any natural material, for example sewage, can be called a pollutant if it occurs in abnormally high volumes and concentrations.

Griffin: One of the problems in discussing the meaning of the word is that the issue is clouded by the sharp division between the environmental activists and informed scientists. We may be dismayed by the publicity the former group gains but if we could only get together, with the aid of educationalists, so that learning and dynamic discussion could take place, we might get a more balanced view of 'pollution'. This is to my mind very much a sociological term.

Sharratt: One must keep an open mind all the time and discussions with the activists are often made difficult because it appears that some have already made up their minds. My view is that pollution becomes a problem when the quality of life is diminished. If this is accepted, is there any evidence that agrochemicals are pollutants? I would say no, they have improved the quality of life. We must also realise we are living in a chemical age and agricultural chemicals form only a fraction of the total problem. A hundred years ago, one had open sewers and so on and people died of disease and malnutrition. That is what I call pollution. Perhaps chemical pollution is more sophisticated and less apparent to the eye, and I think a lot of our fear is bred through ignorance which must be counterbalanced by proper teaching.

Knights: How does one get proper teaching and a complete dialogue when the main communication media have to sell for a profit and good news is no news?

Bagnall: Also, well-known and influential people often start off scares by broadcasting their views before all the facts are known and later discoveries and work which may cast doubts on these views get no publicity.

Major: *British Agrochemicals Association.* What solution will satisfy the environmentalists? If we ban the use of agrochemicals, many people will die of malnutrition and disease. At some time we have to decide whether to allow a certain level of pollution or allow the population die back.

Griffin: In relation to communication, many active environmental groups in the USA are so impoverished, they cannot possibly afford a symposium like this one. It may be idealistic, but we could increase the chances of a meaningful dialogue and help overcome the polarisation of views if one could subsidise the activist to attend. It would also be useful to consult behavioural scientists on the methodologies available to stimulate and allow a better degree of communication.

Miss Hirst: I would imagine the symposium fees for most of the audience have been paid for by their firms or departments, whereas some of us here are paying for ourselves and I know of many other people who can't afford to come. I feel you are all taking a very patronising attitude towards non-scientific environmentalists.

Knights: Does the scientific content of a symposium such as this also discourage the non-scientific environmentalists?

Bates: The scientist has very little chance of reaching the average man in the street. I think this dialogue must be conducted via the people who control the mass media.

Bagnall: The pesticide industry has itself tried to improve communications and a few years ago did launch a Code of Practice. Many reporters were invited to an inaugural meeting but the publicity achieved was pitifully small.

Also, commercial firms suffer in the production of films, literature, and so on explaining their case, because people are cynical and say they are only doing this to salve their consciences.

Mantle: There is a feeling that firms have unlimited funds for publicity. This is wrong, because agrochemical firms either make fairly small profits or exist because the firm has more profitable divisions. The total market in this country is only worth perhaps £30 million.

There is a great mistrust by people of power structures and the possibility that profit motives and political issues can distort the picture. Education on pollution would be best coming from educational establishments.

Reay: In education, it often appears that the student must be deconditioned regarding popularly held views before he can make a balanced judgment. I talked to a university biological society last year and the audience was shocked when I said I believed DDT could be very useful and in fact did a lot of good. They had prejudged the case and some were quite horrified at my views.

Mantle: The ideas of 'vested interests' and 'profit motives' can be carried too far. From the very outset, agrochemical companies have had medical, toxicological and ecological departments looking into pollution problems. To remain in business, agrochemical research and production must be run properly and conscientiously.

Major: Agrochemicals are under constant expert surveillance and I feel one must, as in any other technology, put one's trust in the experts in each field.

Gerrard: *Wolverhampton Technical Teachers College.* This is very fine but the general populace is, I feel, worried most by the remoteness of the pollution problem. One can refuse drugs but one has to eat food, which may contain chemicals. Poisoning of animals and plants, perhaps more so than the poisoning of humans, is a very emotive subject. Without firm evidence and guidance from the scientific community, people are more prepared to accept alarmist views.

Sharratt: This is where education is so important. The fact that people from industry, the government and scientific research organisations have come to this Symposium shows that they all accept the need for continued effort in this area.

Knights: In conclusion then, it appears that many people here feel that education has a very important role to play in producing a more balanced picture of pollution and the use of agrochemicals. It is very interesting in this context that many polytechnics and universities are instituting courses along these lines and in fact, The Polytechnic of North London is planning to inaugurate a course in pollution control, to which this Symposium is a prelude. Perhaps we can hope that education along these lines will help us to get a more balanced picture and to be better able to foresee and deal with potential pollution problems.

Index